Buying for the Future

D1155608

Buying for the Future

Contract Management and the Environmental Challenge

Kevin Lyons

 Pluto Press

LONDON • STERLING, VIRGINIA

 in association with WWF

First published 2000 by Pluto Press
345 Archway Road, London N6 5AA
and 22883 Quicksilver Drive, Sterling, VA 20166-2012, USA

Published in association with WWF-UK
WWF-UK is a Registered Charity, number 201707
Panda device © 1986 WWF
In the US and Canada, World Wide Fund for Nature
is known as World Wildlife Fund

British Library Cataloguing in Publication Data
A catalogue record for this book is available from the British Library

ISBN 0 7453 1346 9 hbk

Library of Congress Cataloging in Publication Data
Lyons, Kevin J.
  Buying for the future : contract management and the environmental challenge /
Kevin Lyons.
        p.   cm.
  Includes bibliographical references (p.   ).
  ISBN 0–7453–1346–9
  1. Contracts—Management. 2. Industrial procurement. I. Title.
  HD2365.L96 2000   99–37471
  658.7′23–dc21   CIP

Designed and produced for Pluto Press by
Chase Production Services, Chadlington, OX7 3LN
Typeset from disk by Marina Typesetting, Minsk, Belarus
Printed in the EU by T. J. International, Padstow

Contents

Preface

Any book that can show a large organization how to make a saving of 2.4 percent of its annual turnover, and at the same time reduce its environmental impact, has to be worth the time it takes to read.

By linking the practice and philosophy of economic efficiency with sustainable development this book demonstrates principles that can apply equally to universities, local authorities, health trusts and the private sector. Economic efficiency addresses budget management and sustainability addresses the ability of any large organization to reduce its environmental impact.

Kevin Lyons is sharing tried and tested tools for managers who are developing Best Value strategies by describing how he has developed links between Rutgers University's procurement, and environmental policies.

He demonstrates that large organizations can exert much influence through their spending power. The notion of leverage gained through large-scale procurement is well established: supermarkets do it all the time. But the principle of using economic clout to achieve sustainable development is not yet embedded in the public sector.

WWF has a vested interest in promoting ideas and proven practice that can show a win to the environment. But this book goes much further: sustainability can be improved from budgetary and operational efficiency. WWF recognizes that this book provides an essential new resource for creative practitioners in many fields, and in all large organizations.

Ken Webster
Senior Education Officer – Community and Governance
WWF-UK

Introduction
The Power of Purchasing

Greening the supplier chain is the next urgent step for corporate environmental responsibility, as consumers, shareholders and the general public raise their expectations. Green credentials can only be claimed for products and services if a clean bill of health can be shown for their whole life cycle, from creation to ultimate disposal, from cradle to grave. At the same time, budgetary constraints are putting pressure on a wide range of institutions to consider the cost implications of current, and any new, practices including environmental management.

Governments too are exerting pressure from above through legislation and regulation on an ever-increasing number of environmental fronts. On whose shoulders will these new responsibilities fall? Where does capacity exist within institutions to handle these new tasks? Whose remit can be extended to embrace environmental concerns?

This book proposes that purchasing professionals (which can include anyone undertaking major purchasing for any organization) are ideally placed to take up this new role and responsibility, particularly in balancing better environmental management, social responsibility and powerful budgetary controls. This conclusion is based on my personal experience over the past ten years as a purchasing officer at a large US State University in New Jersey – Rutgers University. Although my personal experience is very specific, wider international contacts over the years have persuaded me that it has a broader relevance, and it is that encouragement which has led to this book.

Purchasing professionals have been identified for this central role because they deal with resources on a major scale: they are responsible for the acquisition of raw materials and of manufactured products and also, most importantly, they deal with their disposal. This book shows how purchasing professionals can create contracts that include those environmental dimensions requested by consumers and shareholders and required by governments, at the same time as maintaining or even reducing costs. It shows that, both logically and strategically, they are our 'environmental gatekeepers'.

This potential for incorporating wider values into the purchasing process has been recognized in one particular sector in the UK, where a new system for local government purchasing is being introduced. Best Value is replacing the former Compulsory Competitive Tendering (CCT) system in order to enable local government to make purchasing decisions based on their general effectiveness, including – but not only focusing on – cost-effectiveness. The most significant change is the introduction of several preliminary stages prior to the purchasing or tendering procedure, during which local authorities will look more widely and in depth at their overall duties to their communities, and will then identify the priorities for their policies and their strategies for implementing them. For example, 'Local authorities' performance plans need to reflect an authority's corporate objectives such as that of sustainable development' (DETR 1998). Local authorities' commitment to sustainable development is clear from the fact that over 70 percent of UK local authorities are preparing a Local Agenda 21 (LGMB 1997): Local Agenda 21 is the local version of the international agenda for sustainable development in the twenty-first century agreed by world leaders at the United Nations Conference on Environment and Development held in Rio in 1992.

Why purchasing?

Purchasing is a powerful, universally practised system in organizations. There are professional purchasing offices in schools, businesses, hospitals, in government departments and in non-governmental organizations (NGOs), as well as in commercial and industrial organizations. You name it … if an organization wants it, there are purchasing professionals who can find it and negotiate a contract to buy it. And they spend a lot of money doing their purchasing. Already between 50 percent and 80 percent of the annual budget spent in any US company, particularly in manufacturing and service industries and local and national government, is supervised by purchasing departments and the budget share is growing as the specialist skills, knowledge and experience of purchasing officers become increasingly in demand.

For many years, purchasing processes or systems have made significant contributions to economic stability and protection. Legal contracts protect both the buyer and the supplier and provide a secure base for trading.

Buying in bulk through a central purchasing office, or through teams and co-operatives of several organizations, allows discounts for the buyer and an enlarged and guaranteed market for the supplier. Quality and reliability checks protect the consumer. The search for low-cost supplies keeps prices down and company profitability up.

However, professional purchasing has not yet embedded the total cost of operating life-cycle, environmental and sustainable development criteria in its contracts and in the purchasing process. The concept of 'full cost accounting' focuses on this gap and can be defined as 'the integration of an entity's internal costs with the external costs relating to the impacts of the entity's activities, operations, products and/or services on the environment' (Birkin and Woodward 1997).

In 1988, I searched high and low looking for comprehensive, environmentally responsible purchasing guidelines (I'll be honest ... to plagiarize). I did not find one. For me, even today, this whole process is still an adventure in a foreign land; there are some guidelines out there, but they are only words on paper. The words have to be converted into actions. In the late 1980s, environmental laws were knocking on our doors and no one was home. In an effort to do what I thought was right, I began to experiment with merging environmental initiatives into the established purchasing system.

Purchasing and the environment

My early thinking on the union between purchasing and the environment was simple. Is the purchasing process ignoring fundamental social and environmental issues? If so, what significant and immediate modifications could and should be made to the existing procurement process? What changes were needed and how could they be made without jeopardizing the needs of the purchasing institution? How could a purchasing agent make significant environmental enhancements to the procurement process? Could the purchasing process perhaps become a powerful tool to institutionalize environmental preservation and sustainable development practices? Above all, could new purchasing initiatives based on environmental principles be significantly more cost- and resource-effective for the university? If we could demonstrate substantial cost savings, we would certainly catch the attention of our superiors and win them round to the new initiatives. These savings can now be demonstrated: at Rutgers we have shown that purchasing based on environmental principles can save 2.4 percent of costs, year on year (see below). However, in the early days, that evidence did not exist.

One thing was clear from the start: the purchasing contract was the key instrument. Through the power of the negotiated competitive contract, the regular process of hammering out agreements with suppliers and contractors became the ideal arena for implementing a variety of environmental programs.

Purchasing and waste disposal

It is not just 'buying in' that we are talking about. We need to consider where our purchases go when we don't want them any more. We should be taking full ownership of our 'stuff' from start to finish. The products that we procure travel to us to be used. However, when most of our products outlive their useful purpose for us, they continue their journey to another user or to a landfill or incineration system. This waste disposal area could be near you ... or near someone else! In 1996, New Jersey, one of the most densely populated states in the US, disposed of approximately 30 million tons of waste.

At one time, that waste was often created from a new product and its packaging material. These items travelled through procurement processes, their waste potential unremarked and unconsidered. Now, more often, waste goes to a recycling center or, very rarely, goes for re-manufacturing or de-manufacturing; all systems through which new products can be created to be purchased again, in some other form. As a purchasing professional, I am aware that most of my business transactions, through the usual purchasing process, end up as part of the monuments of solid, and sometimes hazardous, waste problems that we all have to deal with on a daily basis. Until this trend is modified or changed, I am also aware that these environmental problems will continue. Even worse, on any given day, I can even smell some of my efforts; Rutgers-Camden and Rutgers-New Brunswick each have one landfill within a couple of miles from each campus, with Camden having the added bonus of an incinerator. Somehow this 'it's in my backyard (IMBY)' disposal situation almost seems ironic; 'you bought it so now you can live with it ... daily!'

So purchasing involves more than acquiring new products and services. It also involves supervising waste disposal of all kinds, including toxic and hazardous wastes.

How can we become 'gatekeepers' for the environment?

To become 'gatekeepers', purchasing professionals would go to work and perform their normal purchasing functions in the way that they normally do. However, in an ideal world, they would only execute contracts which met the following criteria: that purchases and services bought were not only cost competitive and of high quality, but were also environmentally sensitive and responsible.

This new way of working would also have to consider the environmental impact of global purchasing *processes*, as well as products and services. Purchasers would have to understand the many ways in which services such as heat, light, water, transport are delivered or used and how their purchase and use affect the environment. The ultimate goal would be the creation of a program to design sustainable policies, programs and practices. Such a program would, in turn, guide the purchasing profession (and all people who purchase) down the road towards effective, universally practised, environmentally sensitive purchasing actions in all sectors, including by industry, health care, educational systems, government and by ordinary private consumers. Once this was accomplished, this way of working and of thinking would become part of standard operating procedures. Articles could be written, for trade papers and for research journals, to spread the word. It sounds ambitious, it sounds impossible, but I believe that it could be done and, furthermore, I believe that it *must* be done.

Since 1988, we at Rutgers University have conducted research into the purchasing process to see if this universal practice holds the key to environmental preservation, sustainable development and best value practice. Over the last nine years, we have worked to integrate environmental concerns with the university contract and procurement system. We have incorporated a holistic approach on this city-sized campus, which serves 60,000 students and 9,000 faculty, to meet a goal fostering sustainable development. We have made continued efforts to ensure that all contracts are environmentally responsible and to reduce and eventually to eliminate the greater part of the solid waste disposed of by the university. The many different contracts and programs developed at Rutgers University require the best environmentally sensitive techniques and performance from our contractors and suppliers, and also from our university community. All this has been done at the same time as reducing overall operating cost and waste management cost.

By taking prudent procurement and operational sustainable development initiatives through the procurement process (including cross-organizational input, assessments, research, contract specification enhancements and supplier negotiations), environmental protection measures have been realized as well as reducing costs. By reinvesting these savings into further energy and sustainable contracting initiatives, the savings 'revenue stream' has and will become large enough to significantly reduce deferred maintenance, operational costs and the university's ecological footprint (see Wackernagel 1996). For the period 1987 to 1998, Rutgers University has been able to reduce its operational and energy costs by 2.4 percent using our Environmental Contract Management/Sustainable Development system, in co-operation with our Facilities Maintenance Division and the many organizational stakeholders throughout the university. Major areas of savings have included:

- Lighting. Rutgers has initiated an energy conservation program to replace existing building lighting with more efficient lighting fixtures in 30 major buildings on the Newark and New Brunswick campuses, making an anticipated annual saving of $869,000.
- Energy management computer services (EMCS) have been installed to control building heating, ventilation and air conditioning in major buildings on Newark, College Avenue and Cook/Douglass campuses, which will save in the region of $380,600 through improved management of building environmental control systems.
- Solid waste recycling. During 1988–97, it cost Rutgers an average of $141 to dispose of a ton of waste, including collection and landfill costs. Collection and disposal of recycled materials costs about $26 per ton, saving $115 for every ton of solid waste that is recycled. In 1993, Rutgers recycled 16,187 tons of solid waste, saving $1,861,500 compared with landfilling.

  Parallel to the waste management program, procurement programs are in place to return packaging and recyclable products to contracted vendors thereby eliminating packaging and commodity waste from the university's waste stream altogether. The contracted waste haulers are also required to participate in environmental and social education programs on campus and are required to identify and report any market trends and ideas that will enhance Rutgers' recycling and waste prevention efforts.

Is it really your responsibility?

A purchasing official is in fact a 'professional consumer' and has a responsibility to do what is in the best interests of their organization. If environmental regulations or policies exist within your region, you should be playing an active role in executing these requirements – it is the law in most areas of the world. More immediately, becoming involved is a 'value-added' benefit both to you and to your organization. More and more companies are looking for individuals who can provide additional benefits to the company or organization. Environmental preservation and sustainability practices are raising their heads everywhere in society, and purchasing is not excluded. You would certainly stand head and shoulders above the average purchasing professional with the addition of environmental gatekeeping as an entry on your résumé. Becoming involved in environmentally responsible purchasing programs greatly enhances your career progression.

Purchasing is a powerful process: it controls many resources, enacts policies and implements legal requirements. Being in such a strong position brings both rights and responsibilities in relation to our own organizations and to the wider world. The addition of environmental preservation and sustainable development criteria can enhance the purchasing process in many positive ways as well as making a major contribution to achieving sustainable development.

Environmental responsibility: the broader picture

We all wear many different hats: we inhabit different roles and carry a variety of responsibilities. We are members of society, of communities, of families: we are not solely professional people earning our living. Like most of you, when social issues are the main topic, I am one of those whose mind is never at rest. I always have that simple plan for solving the world's problems ... but lack the skills or resources to do anything about them. For the last 15 years or so, my mind has been particularly focused on a few issues which never seem to go away: over-consumption and waste, and the movement towards globalization.

We are moving extremely fast in these modern times, changing as much in one year as we did over seven in the past. The resources we draw on to support the ever-increasing demands of this lifestyle are tied to the very much slower biological system of regrowth, regeneration and absorption, which works over a hundred-year, a thousand-year, or an even longer timespan.

I believe the speed and spread of globalization has intensified and accelerated the problem. As a purchasing professional, I have ordered and bought just about every commodity known to mankind. Through my daily purchasing activities, I have first-hand knowledge of the strain the types of contracts and the commodities we buy are putting on our natural resources, in our own country and worldwide. I know where those natural resources are located and how to negotiate a contract to buy products which contain those resources. I have those products packaged and shipped to my organization in a variety of ways. Finally, as a purchasing professional, when resources close to hand are used up, I can search the earth (searching even faster now with the advent of the computer) to find alternative resources that have not been tapped into yet. But now I have come to believe something new: that as a purchasing professional, one can have the wisdom to conduct these purchasing activities in a more environmentally responsible way.

Practical, powerful and professional ways to sustainability

Environmental preservation was one of the first global issues that I believed I could actually do something about. At first, the environmental movement was a hazy subject for me. Who was doing what? What impact were their actions actually having? When I started reading about environmental issues, I could only judge their likely cause and effect if they related closely to my familiar daily activities as a purchasing officer. For example, I could not see what impact environmental laws or environmentalists were having on global issues. When first published, environmental laws and regulations looked as if they would be strong and effective, but once business leaders and politicians started crying about them, it seemed as if legislators would have their power defeated by stubborn resistance from those whose profits depended on ever-increasing consumption. In addition, I discovered that no one was fully complying with the environmental initiatives anyway, so why should I bother? Even when I looked at my own organization, it appeared that not everyone seemed interested or concerned. I concluded that perhaps I should leave the politicians, environmentalists and business leaders to work it out.

Finally, the reality of environmental requirements came crashing down on the citizens of New Jersey and on to my desk. New Jersey state officials had passed an aggressive and enforceable recycling and source reduction law and Rutgers University (being a state university) had to comply. It was at this point, between 1987 and 1988, that I began an individual crusade to make a contribution to solving this environmental problem. By bringing environmental issues into an existing, well-established, universally practised profession, purchasing and procurement, we would be involving a group which had not yet done much to contribute to environmental solutions.

In addition, I concluded that the laboratory for this activity should be a well-respected institution of higher learning, because one of its missions is to educate the next generation of citizens and the next generation of leaders. Education-by-doing has always been an effective way to help people accept and learn new initiatives and I believed the university would accept this concept for itself as well as for its members. If the Purchasing

Department at Rutgers University could participate actively in discovering solutions for environmental preservation, we could have important findings to share with students, faculty and staff, and the citizens of New Jersey. This book covers the ten years of personal environmental action, from 1988 to 1998, a ten-year period that has been exciting as well as frustrating and challenging. The actions designed and accomplished during that time are now summarized in this book so that others may make use of them.

The shape of this book

Let me describe the shape of this book. The first four chapters are very practical and down to earth. Chapter 1 takes us through the purchasing process and shows the changes that will need to be made if environmental elements are to be built into future contracts. Chapter 2 tells the story of how environmental factors were built into the everyday operations of Rutgers University, and how these linked to the local community of suppliers and contractors. Chapter 3 looks at how to 'institutionalize' these changes and build them permanently into the regular functioning of the organization. Chapter 4 gets down to the detail of contract design. Chapter 5 shows us some of these contracts in action, in particular what purchasing can learn from the all-important realm of waste disposal. We see how thinking about the end destination of products before we have even bought them can lead to major cost savings.

In Chapter 6, we look at the type of relationships we can develop which will support our own efforts both within and outside our own institutions, particularly for academic bodies working with students, with international academic networks and through the Internet, and allow us to learn from them the best practice elsewhere. Chapter 7 gives some examples of the impact of environmentalism on some major corporations and multinational companies, and how we can work with them. Finally in Chapter 8 we look at what is going on in communities, towns and cities where local communities in the US, the UK and Australia are meeting the environmental challenge, and how their innovative approaches can inspire purchasing professionals and others to work with them to ensure that the local community and the environment can benefit from our new sustainable practices as well as the plan as a whole. At the end of the book, you will find suggestions for further reading, and contact addresses for helpful organizations (including on the World Wide Web).

Finally, to return to the message of this introduction – what can purchasing do to resolve environmental dilemmas? The answer to this question is where all the fun lies and where my story starts.

1
The Purchasing Process

Setting the scene: a guide to the purchasing operation

Purchasing professionals create and authorize contractual agreements with many different vendors for multiple goods and services. They keep pace with the surrounding world and know what types of products and services to bring into their organization to keep it operational and thriving. In short, they know the pulse of the organization and orchestrate the flow of goods and services into the organization.

Are you a purchaser? The US Bureau of Labor Statistics (1998) describes the differences between purchasing and buying as:

> Purchasers and buyers seek to obtain the highest quality merchandise at the lowest possible price for their employers. In general, purchasers buy goods and services for the use of their company or organization and buyers buy items for resale.

The Bureau suggests that the various tasks and functions of a purchaser include the following:

> They determine which commodities or services are best, choose the suppliers of the product or service, negotiate the lowest price, and award contracts that ensure that the correct amount of the product or service is received at the appropriate time. In order to accomplish these tasks successfully, purchasers and buyers study sales records and inventory levels of current stock, identify domestic and foreign suppliers and keep abreast of changes affecting both the supply of and demand for products and materials for which they are responsible.

Purchasers find information from many sources to inform their decisions:

Purchasers and buyers evaluate and select suppliers based upon price, quality, availability, reliability and selection. They review listings in catalogs, industry periodicals, directories and trade journals, research the reputation and history of the suppliers, and advertise anticipated purchase actions in order to solicit bids from suppliers. Meetings, trade shows, conferences and visits to suppliers' plants and distribution centers also provide opportunities for purchasers and buyers to examine products, assess a supplier's production and distribution capabilities, as well as discuss other technical and business considerations that bear on the purchase. Specific job duties and responsibilities vary with the type of commodities or services to be purchased and the employer.

How can the role of purchasing impact down the line, down the supplier chain? How might it even have influence on product development? And how does purchasing impact on other sectors of business and industry?

Changing business practices have altered the traditional roles of purchasing professionals. Manufacturing companies have begun to recognize the importance of purchasing professionals and increasingly involve them at most stages of product development. Their ability to forecast the cost of a part or materials, availability, and suitability for its intended purpose can affect the entire product design. For example, potential problems with the supply of materials may be avoided by consulting the purchasing department in the early stages of product design.

Increasingly, purchasing professionals work closely with other employees in their own organization when deciding on purchases, an arrangement sometimes called team buying. For example, they may discuss the design of custom-made products with company design engineers, quality problems in purchased goods with quality assurance engineers and production supervisors, or shipment problems with managers in the receiving department before submitting an order.

How many purchasing professionals are there in the United States?

Purchasers and buyers held about 639,000 US jobs in 1996. Purchasing agents and purchasing managers each accounted for slightly more than one-third of the total while buyers accounted for the remainder. About one-half of all buyers and purchasers worked in wholesale and retail trade establishments such as grocery or department stores. One-fourth worked in manufacturing.

Purchasing professionals do not operate alone

Purchasing is a massive system. The information given below reveals that purchasing actions may involve just about anyone or everyone in an organization, whether they have the title of purchasing officer or not.

According to Claudia Montague's study for the Center for Advanced Purchasing Studies (CAPS):

> Purchasing departments may handle almost half of the dollar amount of purchases made by business and government. Overall, CAPS found that purchasing departments handle about 41 percent of an organization's total expenditures. The rest of the money is being spent by everyone from corporate officers to department heads. (Montague 1995)

Things are changing: Montague's report concludes by quoting an earlier CAPS study of purchasing executives which stated that '87 percent believe buying *teams* will be making sourcing decisions in the year 2000.'

Based on this, it is clear that many people have purchasing experience, so we have a great starting point and mutual understanding for what we can do to enhance this process.

The preliminary steps to designing contracts

Prior to executing contracts, there are multiple steps that purchasers need to take. The steps we take to design a contract is where all the hard work, the regulations and procedures behind the purchasing process, have thrived for many years. This is serious business and, at times, not too much fun. However, hidden within all this starchy, hardline contract preparation process is where all the fun for me has been for the last ten years. Looking within and beyond the obvious purchasing policies, procedures and guidelines, and breaking into the established system to find the tools to institute environmental preservation schemes, has been a non-stop joyride. However, there is a warning that goes with this type of work: once you begin this exciting process, it is hard to stop.

Do you know enough to get involved?

If you would like to try to become an 'environmental gatekeeper', you will not need special training or certifications to master this new and important responsibility. Environmental gatekeepers are individuals who want to make a positive environmental impact on their organization and the surrounding community, to preserve historical economic and quality assurance practices and goals, and to maintain the competitive business ethic while performing the task of professional purchasing.

Firstly, examine the existing purchasing processes you currently work within. Can you identify where your products and services are coming from and what impacts they are having on your organization? This identification process should start with specific priority areas, set by policy discussions and consultations with key officials in your organization. Then you can begin to identify the resources used to manufacture the products in these key areas and to consider where they will go after they have served their useful life.

Getting started

To enter into the exciting and adventurous world of environmental gatekeeping and executing sound life-cycle or sustainability practices you must begin by asking some basic questions from a purchaser's perspective.

What is environmental preservation and how is it connected to purchasing?

Purchasing is primarily responsible for the import and export of goods and services for the organization. The impact these goods and services have on the environment is significant. Buying products which have been made using fewer non-renewable natural resources helps preserve the environment. An effective purchasing contract can dictate what products are made from and how they are made, packaged and shipped, utilized and disposed of. Criteria to deal with environmental preservation are critical and should be contained in the contract.

This is where the excitement begins. Merging traditional purchasing procedures with environmental protection tactics is delicate surgery. First and foremost, the primary goals of the competitive contract process cannot be jeopardized, nor can the standard purchasing procedures and policies be abandoned. However, once you have gone over the obvious obstacles and you begin to dig deeper, you will find incredible solutions and your contracts will have long-lasting environmental benefits to your organization.

What government regulations support environmental purchasing activities?

If you check the environmental regulations you need to abide by at every level (from local to international), you will find specific guidelines for establishing environmental performance and action standards, but you may not find step-by-step guidelines on how to incorporate these initiatives into your established purchasing systems and negotiated contracts. These standards must be incorporated within the purchasing process in order to comply with the environmental regulations while, at the same time, remaining flexible enough to change if future regulations are introduced. Issuing environmental regulations and carrying out their actions through the power of purchasing is rare. However, this process holds the key to the development of effective environmental programs. Once again you have set yourself up for a challenge – to understand environmental policy, develop a plan and master the technique of merging environmental policy within your established purchasing program. If you succeed, you will be the departmental hero and perhaps even influence the whole purchasing profession.

Whose duty is it to implement environmental laws and regulations?

In New Jersey, USA, for example, environmental policies and regulations have been developed to set the stage for preserving resources throughout the entire state. However, only a few individuals are responsible for interpreting and practising these requirements and merging them into the operations of the organization. My investigations showed that most individuals with environmental compliance among their duties do *not* work in a purchasing or business division. Therefore, the effectiveness of many environmental programs is short-lived and not maintained because it falls to the wrong people to implement them.

Now for purchasing ... values

The act of purchasing and negotiating contracts for goods and services is an established practice in business. As a purchasing agent or purchaser, you are one of the most significant members of your organization. In most cases, the purchaser has been given the sole responsibility and power to negotiate and make the business decision. You are 'minding the store' and you are, therefore, the natural 'gatekeeper'. And when you are minding the store, your responsibilities are not one-dimensional. You must develop an all-encompassing, holistic approach to gatekeeping. And you must take ownership of your professional decisions: the results of your contractual decisions could have long-lasting implications.

The extent to which purchasing and contracts can be used to implement the values of an organization was one of the first questions to consider in the early days of my environmental work. Historically, goods and services are acquired to support, enhance or match the values and the operational and social mission of the organization. However, the new thinking had to define the values afresh and answer these questions:

- What are the environmental, societal and economic values of the organization?
- How can the purchasing process and purchasing contract assist in institutionalizing and sustaining these values?
- Can I, as a purchasing agent and gatekeeper, protect the values of the organization and at the same time maintain a fair and competitive position?

The tool or instrument that is used to protect the organization is the purchasing contract. Historically, purchasing set economy and quality as its twin goals. These two elements are critical and must form the backbone of any purchasing process. But this backbone of economy and quality can carry a larger load. Environmental preservation, sustainable practices and localized social and economic development initiatives are values which can be built into the official purchasing process. These additional goals and values will reap tremendous benefits both for your organization and for your community. People's livelihoods and local commerce can both be stimulated by appropriate purchasing contracts.

Environmental laws: aid or obstacle to the purchasing system?

Environmental laws in the US have always been met with immediate opposition from industry as a expensive addition to 'business as usual'. During the early part of 1997, the US Environmental Protection Agency tried to introduce new cleaner air standards and industry went crazy. There were even suggestions that the backyard barbecue was in jeopardy.

The competitive purchasing system is a legal process. The documents used are designed to protect you and your organization from any legal problems and to ensure that the organization acquires the best products and services for the best price. Environmental laws are also legal initiatives. They contain criteria and goals for the environmental protection of your organization and your community. However, these laws do not spell out what they will cost to implement. Ordinary citizens and the business community are left to figure out the economic impact for themselves. So how can these two legal systems be compatible when they have such different goals? If you considered the strong opposition to environmental laws from industry, you would have to expect there could never be compromise, no reconciling of such opposing aims.

Since 1987, a phenomenal number of environmental laws have been issued in the US by the Federal Government and by New Jersey's Department of Environmental Protection. One of the most recent federal initiatives is the US Federal Government Executive Order (EO) number 13101, *Greening the Government through Waste Prevention, Recycling and Federal Acquisition*, which was signed by President Clinton in October 1998 (OFEE 1998). This enables the government to comply with multiple eco-procurement initiatives although, as with all federal executive orders, it does not tell you how to accomplish the goals. Nevertheless, it is now clear that the US Federal Government sees just how crucial purchasing is in the scheme of things, and recognize its role as potential 'environmental gatekeeper'. EO 13101 and Local Agenda 21 can be seen as two distinct and complementary US and UK mandates that this book can build on to begin to help stakeholders embrace a new direction of economic and environmental responsibility.

At state level, some of New Jersey's new legal requirements include:

- a target for New Jersey to recycle 65 percent of solid waste stream
- solid waste reduction through source reduction and recycling
- solid waste tonnage and revenue reports required
- Treasury-leased/owned buildings to provide source separation and recycling
- janitor contracts to include recycling collection
- purchase of recycled paper to have priority over virgin one
- a revolving fund of profits from recycling to be reinvested into recycling programs, 10 percent of fund to be spent on education and technical training.

In several cases, it was purchasing agents who were given the role and responsibility for implementing environmental action, including to:

- appoint a co-ordinator from agency procurement staff
- issue Comprehensive Procurement Guidelines (CPG) and Recovery of Materials Advice Notes (RMAN)

- establish affirmative action to purchase products identified by the EPA
- use compulsory guidelines for federal government purchasing departments (OFEE 1998).

Environmental legislation looks like an obstacle to most of us

As environmental policies or procedures are issued, modified or changed, the purchasing process has to follow through at speed with changes which may not have been well researched or tested for effectiveness. Environmental policies have not been friendly to some purchasing systems. The technology or system has not always been fully developed for purchasing agents to negotiate or to execute successfully those new types of contracts which will meet environmental laws. In addition, the purchasing agent must still consider whether contracts are competitive and are within the established financial budget of the organization.

To make matters worse, there are always only limited resources for the purchasing operation to put into researching and benchmarking the advent of environmental laws within the purchasing process. For higher compliance ratings, future environmental laws and policies should contain provisions for assisting purchasing professionals in designing new environmental specifications and contract language. Additional obstacles exist when the way the environmental laws will operate in real-life settings are not fully understood.

Building in compliance – and seeking out the obstacles and benefits

The organization should work towards the careful inclusion of environmental law statements and goals in their competitive purchasing process. The key word here is 'competitive'. If the environmental laws are being used to assist the purchasing process, they must be viewed as value-added/cost avoidance mechanisms. Instead of looking for the disadvantages and difficulties of adding in environmental criteria, look for the advantages that the organization can derive from the new initiatives.

The first and most obvious advantage is to be ahead of the game, to be first with the solutions that others are struggling towards. The second advantage is to build in compliance to the regulations throughout your whole supply chain, thus distributing and sharing the responsibility more widely. The products and services you purchase can arrive 'clean and green' if your suppliers have already complied with the law.

2
Rutgers University, USA

This section tells, step-by-step, the story of how environmental factors were built in to the everyday operations of Rutgers University and, from there, out into the local community of suppliers and contractors, all through purchasing department initiatives.

Rutgers University is situated in New Jersey, one of the most densely populated states in the United States. It is a land grant public university with a population of 60,000 students, 17,500 faculty and staff, and over 650 buildings covering three campuses and five municipalities. Land grant universities are state educational institutions given land by the federal government under the provision of the Morill Act of 1862 on condition that they offer courses in agriculture and the mechanical arts.

Rutgers University is part of five communities that surround the three university campuses. Additionally, Rutgers' population of staff and students is made up of individual citizens who live, shop, consume and dispose of products within New Jersey and the surrounding states. This inter-relationship is important because the examples that we set as individuals and purchasing professionals are transferred into the community.

The New Jersey environmental laws are strict and there are many of them, but when they are utilized to the advantage of the organization, they prove to be the best allies that Rutgers University could have in developing its own environmental policies and programs.

Environmental blending, mixing, action

The design of environmental policy is a crucial part of any environmental program. However, not much attention is paid to how the policy will be implemented. We had three goals in mind when we developed the environmental policies at Rutgers University:

1 To design and introduce policies which match the university mission
2 To establish policies which environmentally enhanced university operations and educational programs
3 To implement the program using evaluation criteria which would ensure that the policies would not remain 'great ideas', but would be policy initiatives that would be implemented and practised.

When in 1987, the New Jersey state government instituted the New Jersey Source Separation and Recycling Act (PL 1987, C102), the lives of all the citizens in New Jersey were immediately affected. There had already been many voluntary initiatives to start various community-based recycling programs (for example, a 1970 student-led recycling program had a short life at Rutgers), but it was the 1987 Recycling Act which launched the mandatory, state-wide recycling programs. The 1987 Recycling Act required:

the recycling of solid waste materials for the purpose of reducing the amount of solid waste requiring disposal, conserving valuable resources and energy, and increasing the supply of reusable waste materials for New Jersey industries.

However, this Act (like most legislation) did not provide resources for formal training or information on how to design a purchasing system which would adhere to this law or support the initiative. In short, the Act listed items that could no longer be taken to the landfill, and individual New Jersey organizations were responsible for setting up programs and redesigning or developing new contracts to adhere to this mandate. Prior to the 1987 Act, in Rutgers University, as in the majority of New Jersey organizations, contracts were set up to send all waste to the various New Jersey landfills and other out-of-state disposal facilities – every type of waste thrown together in one container. Since then we have realized the enormous variety of valuable commodities in the New Jersey waste/recyclable stream, including yard waste, food waste, newspaper, corrugated paper, office paper, other paper, plastic containers, other plastic packaging, other plastic scrap, glass containers, other glass, aluminium cans, foils and closures, other aluminium scrap, vehicular batteries, other non-ferrous scrap, tin and bi-metal cans, white goods and sheet iron, junked autos, heavy iron, wood waste, asphalt, concrete and masonry, tires, other municipal and vegetative waste, other bulky construction and demolition waste (list from the New Jersey Materials Specific Recycling Rates 1995).

Moving beyond 'demand reaction' management

Most purchasing systems are set up as 'demand reaction' systems (purchasing reactions to a demand/request). After a request has been made to procure a product or service, the purchasing agent begins the research and execution of the contract. Purchasers rarely stray from this strict process. However, as the recycling laws and other environmental laws started to impact on the purchasing process, there were added pressures. The standard purchasing process had to make room for new guidelines and procedures without altering the straightforward process. Our initial feelings were that we did not have sufficient time to research or test the implications of environmental policies as they were issued.

Six weeks after the Act was signed, Rutgers University had to have a source separation program in place. After the first year of the Act, we would be required to recycle 25 percent of our waste stream. No organization could put paper, cardboard, bottles and cans any longer into the landfill waste stream. The Act did not say where to take this recyclable waste, so we needed to create a program to separate these products at source and prepare them for the recycling process. During this six-week period, officials from the university's Housing, Dining Services, Purchasing, Facilities Maintenance and Environmental Health and Safety departments all met to plan and implement our new waste flow procedures. One of the most immediate concerns was the renegotiation of the existing waste management contract. The contract was renegotiated to meet the immediate criteria of the new law, but everyone agreed that a new comprehensive waste management and recycling contract would have to be designed and competitively bid to replace this renegotiated contract.

In 1988, well after the law became effective, an investigation was commenced and research was conducted into the development and design of a comprehensive, competitive waste management and recycling contract to replace the renegotiated contract. This was the first step which led to the official development of the environmental gatekeeping program. Once again, officials from the university's Housing, Dining, Facilities Maintenance and Environmental Health and Safety departments came together and provided input into the criteria for the new waste management/recycling specifications. Each individual had critical input on what their operation was responsible for and how the new contract could assist them in completing their operational mission. As a purchasing official, it was my responsibility to have these departmental criteria merged into a competitive contract which was compliant with the New Jersey law. It was at this critical point that it became clear that it was possible to go beyond the required environmental regulations and laws, to find ways to enhance the contract and to prepare for future environmental changes – in other words, to be ahead of the game.

Relevant environmental laws and regulations

As well as knowing New Jersey environmental law, some time was spent looking at several environmental laws from around the United States, including those passed by the federal government; this is an ongoing process. Some of the initiatives were similar, and they contained useful language, terms and phrases that could be incorporated into our own contracts. For example, an extract from New Jersey State Executive Order 34, announced by Governor Jim Florio on June 13, 1991 gives a taste of the language, terminology and requirements (note in particular that item 5 stresses the importance of education and training and technical assistance to support the new legislation):

I direct that the Department of Treasury:
1 Ensure that all solid waste contracts written for solid waste collection and disposal include provisions that reflect a reduction in waste generation rates achieved through source reduction and recycling, and provisions that ensure conformance to county and municipal recycling requirements.
2 Require contract vendors to provide tonnage and revenue reports to each agency or instrumentality.
3 Ensure that all contracts for leased and owned buildings under the jurisdiction of the Department of Treasury provide for source separation and recycling programs.
4 Ensure that all janitorial contracts awarded by the state include recycling collection services, where feasible.
5 Establish a revolving fund, consisting of moneys derived from the contracted sale of recyclable materials, to be used in supporting state agency recycling programs. It is further ordered that at least 10 percent of the total fund be allocated to the Office of Recycling in the Department of Environmental Protection to support educational and technical assistance programs needed to develop and maintain state agency recycling programs. Any revenue derived from the sale of recyclable materials sold as part of a training and education program for the rehabilitation of individuals at state institutions may be used to provide direct support for these programs.

Executive Order 91, launched by Governor Jim Florio on May 5, 1993, once again identifies the all-important role of purchasing and procurement departments in implementing the new legislation. This Order also gave very detailed information on proportions of post-consumer waste to be included in recycled products. It then proceeded to detail, in multiple paragraphs, information on the percentages and amounts and other specifications:

> ... each state agency and instrumentality shall appoint within 30 days of effective date of this Order, a co-ordinator from the agency procurement staff who will be responsible for co-ordinating with the Division of Purchase and Property and the Division of Solid Waste Management, the procurement of recycled products by the agency or instrumentality. The co-ordinator's responsibility is to assure agency compliance with the purchasing goals of this Order, etc.
>
> Require the purchase of recycled paper and paper products of comparable quality with virgin products, with consideration given to recycled paper and paper products containing the highest percentage of post-consumer waste paper material, when such purchase is competitive as defined below. In the event that this requirement is impossible to meet due to mill or vendor supplies of paper and paper products containing recycled secondary and post-consumer content, each state agency or instrumentality shall meet, at minimum, the following purchasing schedule ...

The all-important Buy Recycled program was launched by the United States Environmental Protection Agency (USEPA) in 1995. It was no use collecting waste material, turning it into recycled products unless there was a guaranteed market out there waiting to buy those products. This is known as 'closing the loop' and the concept recurs throughout this book as one of the foundation stones of sustainable development. The Buy Recycled program states:

> As part of its continuing program to promote the use of products containing recovered materials, the Environmental Protection Agency (EPA) issues Comprehensive Procurement Guidelines (CPG) and Recovered Materials Advisory Notices (RMAN). The CPG is the basis of the federal government Buy Recycled program and designates products containing recovered materials for government agencies to purchase. The product designations are organized by product category: paper and paper products, vehicular products, construction products, transportation

products, park and recreation products, landscaping products, non-paper office products, and miscellaneous products. The RMANs provide recommendations for purchasing the products designated in the CPG. Along with the CPG and RMANs, EPA publishes technical background documents that provide more detailed information about each of the designated items, eco-purchasing fact sheets, and lists of manufacturers or vendors of the designated products. Through use of the guidelines and related information, the federal government hopes to expand its use of products containing recovered materials and to help develop markets for them in other sectors of the economy.

The Buy Recycled program required 'affirmative procurement' to make it work effectively. Section 6002 of the Resource Conservation and Recovery Act required 'procuring agencies' to establish affirmative procurement programs for purchasing the products designated by EPA. 'Procuring agencies' are federal agencies, state and local agencies using federal funds to purchase the designated products, and government contractors. RMAN 1 provides EPA's recommendations for developing affirmative procurement programs. Additional information is available in *Greening the Government. A Guide to Implementing Executive Order 13101* (OFEE 1998).

State, federal and international laws consistently affect the purchasing operation and how we perform our daily job. When the 1987 the New Jersey Recycling Act was passed, I read the legislation and took careful note of the wording it contained. Purchasing professionals should set aside time to review all the relevant documents thoroughly. Knowing what is contained in these publications can prevent some costly mistakes and can give you a head start on creating your own style of environmentally enhanced contracts, following the lead given by the legal and regulatory framework. Most of the environmental laws are written to invoke 'personal responsibility', which means the environmental policies are meant to be implemented by individuals taking an initiative to act. However, you can either interpret them and institutionalize them in a way that can benefit your organization, or you can do it incompetently and find that you have created obstacles and reduced your ability to comply as well as the effectiveness of your compliance.

Initial criteria

Prior to merging environmental regulations and policies into the purchasing process, you must develop your own policies and be very clear about how your organization operates. This is needed so that you can show

how to implement these laws throughout the entire organization without overlooking procedural concerns or leaving someone out of the plan. To make this program meaningful and sustainable during its developmental stages, we needed some clear steps to help us get started. As a result, in 1988 we developed the following criteria within the purchasing department:

1 Strict adherence to the state regulations and university policy and procedures
2 Environmental procurement strategy to be communicated to all prospective vendors and contractors
3 The environmental changes needed will be communicated to the entire organization
4 The purchasing department to create and implement environmentally sensitive programs and contracts which are departmentally user-friendly and cost-effective using the best value system
5 Seek to exceed all existing national and international environmental and recycling goals, and implement opportunities to achieve this target
6 Reduce the overall volume and cost of waste management and recycled content products via the contract
7 Enhance the contract specifications by seeking multidisciplinary environmental research and technology using university faculty and students from all academic areas
8 Assist in implementing local/state economic development with programs to transfer environmentally sensitive technology
9 Assist local communities, school systems, universities/colleges, governmental agencies and political subdivisions and industry in the development of their environmental and recycling program
10 Utilize the above programs (1–9) to reach the goal of 'Rutgers Investing In Itself for a University-wide Environmentally Self-sufficient and Sustainable System' (extracts from Rutgers' paper Initial Criteria for Environmental Purchasing).

The last four items above demonstrate the wide-ranging and lateral thinking that was developed, such as the idea that the university would freely offer its expertise to contractors, and the idea that contracts would seek to make a positive contribution to local economic development. Both these ideas are developed more fully in later chapters.

In 1988, when we began our research into the contents of current environmental laws, it became clear that the university's standard purchasing procedures and guidelines were in direct conflict with these laws. To execute contracts which complied both with these strict environmental standards and with the university's purchasing procedures, we would have to develop an entirely new process. However, this process would have to involve more than just a few commodities; the environmental laws affected the entire organization and a substantial enhancement of the purchasing process was required. This is where the environmental contract management program began – by looking at environmental requirements through the eyes of a purchasing professional. As our department negotiated contracts for goods and services for the entire organization, we were the weak link which required immediate attention. This was a powerful and important discovery for me because it showed that the weak link could be rethought and transformed into the strong and influential link. The new contract could become the lever and the enhanced purchasing process could act as 'gatekeeper' for the environmental agenda. What we learned enabled us to build in and institutionalize environmental programs and to continue developing new ones as required. The guidelines on how we were able to use the power of the purchasing process is the focus of this book.

The first task: waste

As a purchasing agent responsible for the acquisition of many goods and services, some of my first assignments were to deal with waste management, recycling services, water and sewage and hazardous waste. As a result of multiple federal and New Jersey state environmental regulations, I determined that the specifications for each of these commodities would have to be revised: the purchasing official had to execute the competitive contracts which met environmental policy at the same time as protecting the economic interests of the university. This incredible conflict formed the catalyst for my environmental contract management actions. The steps taken then reflected the environmental, best practice/best value and sustainability ethic that I still use today. These actions are common to all organizations and can be applied to purchasing operations in government, industry, health care and educational institutions.

Historically, purchasing has not been conscious of any responsibility for the protection of the environment or of the environmental performance of a company. Purchasing has often been done with no thought for the way products are made, packaged, shipped and used. Nor has purchasing concerned itself with what happens to those products, packaging and services once we are ready to get rid of them.

We set out our initial dealings with relevant environmental laws/issues in our early waste contracts:

- Disposal of trash must be in compliance with all United States, federal and New Jersey, state, county, city, EPA, DEPE, laws, rules and regulations.
- Rutgers reserves the right:
  - to identify and remove trash items from the general trash stream and have those items targeted to recycling or remanufacturing processes
  - to reduce the scope of this contract with the intent of increasing recycling compliance, new flow control regulations or to support alternate waste management strategies or programs; Rutgers will negotiate these changes with the contract vendor prior to executing the plan changes
  - to select the disposal site for its trash. If/when flow control laws are passed, Rutgers reserves the right to negotiate with the contract waste hauler the best waste receivable site for its trash
  - to tour all facilities that will be used as part of this contract within three months of the award of this contract.
- Waste audits: contract vendor will participate in periodic waste audits which will be conducted on campus. After award of contract, contract vendor will work with Rutgers to plan a three-year waste audit schedule.

The contract included a ground-breaking element: it required the successful contractor to educate the university in recycling:

> Recycling Education Plan: please describe your plan to educate the faculty, staff and students about your waste management/recycling. The goal is to have the entire campus recycling at a rate which is in accordance with NJ State Executive Order Number 34 (60 percent). How will you introduce the program, and how will it be sustainable each year of this contract?

Existing contracted vendors can also provide additional resources for information. In most cases, these vendors are also mandated to comply with environmental laws and must institute these practices as part of doing their own business. Including the vendors' perspective on the environmental laws was an important factor in the design of the Rutgers University contracts. In this case, it was the environmental laws seen from our perspective and also from the vendor's perspective that stimulated some of our contractual actions.

Developing an environmental policy and an environmental purchasing policy

We needed to have a university-wide policy which was part of the university's overall mandated policies procedures. In 1990 a group of students from the Students for Environmental Awareness (SEA) organization, plus the graduate advisor, commenced work on the university-wide environmental policy. With a lot of blood, sweat and tears, they hammered out and passed through the University Senate the first university-wide environmental policy. The two key resolutions in this early document were these:

> Resolved ... to support a policy on recycling
> ... to encourage the formation of a Recycling and Source Reduction Committee (RSRC), composed of students, faculty, and administrators
> ... to encourage appropriate curriculum committees to study the possibility of developing a required core course on Environmental Citizenship (Rutgers 1992a).

In the second point you will see that, from the outset, we planned to put the new environmental agenda into the curriculum, where the real-life actions taken within the university would be underpinned and sustained by academic support.

The next step was to identify the practical ways in which we could carry out our general aims. As an example, the university's environmental policy included a long list of all the items targeting paper use and paper recycling:

- Encourage all units involved in publication of materials to use recycled paper whenever possible and to note the specific make-up of recycled contents.
- Encourage university bookstores and Co-ops to comply with university goals for the offering of recycled content products at comparable or lesser prices than virgin.
- Encourage ReproGraphics Services to offer several grades of recycled paper and to promote the use of recycled paper unless the customer requires virgin.
- Encourage all faculty who have class materials photocopied by off-campus vendors to specify exclusive use of recycled paper, and to avoid use of colored papers which are more expensive to process.
- Encourage the university libraries to load all photocopiers with recycled xerographic paper and to make specific notation of that practice on the equipment.

- Encourage the university libraries, when replacing older photocopiers, to specify that new equipment must accommodate recycled content paper and have the two-sided copying feature.
- Encourage the University Senate to utilize products with recycled content whenever possible.
- Encourage the double-sided copying of all documents which are longer than one page, unless the requester specifies otherwise.
- Encourage the programming of photocopiers with a default setting for double-sided copying where possible.
- Encourage the senders of short memos intended for mass distribution to photocopy the document two or more times per sheet of paper.
- Develop and phase-in computer-aided mailing lists for specifically targeted mass mailings, with the assistance of Computer Services and Personnel Services.
- Encourage the reservation of university bulletin boards for the use of university activities.
- Encourage progress toward collection of additional categories of recyclable paper materials.
- Make mailrooms and copyrooms as recycling-friendly as possible.
- Discourage use of colored or glossy paper.
- Encourage university bookstores to charge $.02 per bag, and to discount purchases over $5.00 by $.02 if no bag is taken (Rutgers 1992b).

There were many topics other than paper covered by the original list, including disposal of household hazardous wastes, design and construction of new residences to be recycling-friendly, food service operations to reduce use of disposable items, discount for use of own cups in cafeterias, phasing out of aerosol cans and – always remembering the need to build in education and awareness – orientation sessions on the program for all new students.

This environmental policy was a major step in the right direction for the university, and also for the purchasing department as most of the items had a purchasing connection to them. At that time students were working with us on multiple environmental issues because we had many goals in common. When the new policies were voted on and passed by the Senate, I began to work with the students so that they could understand the university's purchasing process and see what assistance we could provide in implementing these initiatives. As an additional bonus to the policies being passed, the university president nominated me to chair the Recycling and Source Reduction Committee called for by the policies. One of my first actions was to create a purchasing version of the Environmental Policy, tailoring it to the operations in purchasing. To develop these environmental purchasing policies, all relevant environmental policies and laws were researched and the criteria contained in these documents were merged into our existing purchasing policy and procedures. This was a major undertaking, but the size of the task did not defeat us. Every purchasing system has its own policies and procedures and these *can* be blended together with environmental laws and policies to create a powerful program document.

The environmental purchasing policy program that we created at Rutgers University is not a 'one size fits all' but it is a program which works through the power of the purchasing system. The effective environmental gatekeeper will need an effective environmental purchasing policy to develop and institutionalize an iron-clad program. Rutgers' Environmental Purchasing Policy included the following elements:

- To institutionalize the use of the 'Total Product Life-Cycle' system for evaluating what environmentally sensitive techniques all vendors must use to develop, design, package, ship and return of the various products (and packaging) to vendors for recycling, reuse, remanufacturing. This 'closed-loop' scenario should be the first choice in the design of environmentally sensitive contracts.
- Establish a resource tracking system to trace and monitor the closed loop.
- Enforce the Return/Reuse Packaging Program.
- Utilize the [... existing procurement processes ...] to incorporate innovative procurement strategies.
- Phase out hazardous cleaning products. Assist in the research for high quality, cost-effective, environmentally sensitive alternatives.
- Increase the Contractor Environmental Commitments – to be kept on file.
- Enforce the Contractor Environmental Public Awareness Program via the contract.
- Require *all* contract vendors to identify and notify buying staff of all environmentally sensitive products or services they provide and will provide in the future.
- Develop a return packaging program or environmentally sensitive packaging program. Consult with the Packaging Engineering Division for the latest information and research.
- Establish the use of *all* available university environmental research or technology, developing technology transfer contracts if necessary, in order to add environmentally sensitive language to specifications.
- Design environmentally sensitive specifications.

- Conduct quarterly research into current recycled content product market trends.
- Provide contract assistance to Facilities Design in incorporating energy efficient designs into all specifications, plans and drawings for construction projects.
- Develop Environmentally Sensitive Hazardous Waste Contracts with the assistance of Environmental Health and Safety.
- Expand and incorporate the Economic Development programs as a result of 'closed-loop contracts' (such as through technical transfer projects with the Center for Plastics Recycling and Research (CPRR)).
- Utilize the Procurement and Contracting student internship program for research and policy development, each semester, open to all academic curricula). *Note:* A faculty advisor must be used in conjunction with purchasing acting as the technical advisor.
- Institute a ban on the procurement of any tropical hardwood products or products which were produced using tropical hardwoods unless procured through sustainable managed timber forests (check with the Center for Plastic Recycling Research Division prior to making these types of purchases).
- Procure recycled content paper with specifications which are in accordance with all federal and NJ state guidelines. (Rutgers 1993)

Rutgers Environmental Purchasing Policy document also highlighted the need for education and awareness-raising as an essential ingredient for success:

- Establish an outreach program to assist universities, school boards and government entities in the development of environmentally sensitive programs (help with information and contract development).
- Use the policy to enhance interdisciplinary programs, and the academic environmental literacy programs on campus.
- Develop a university-wide communication and community information program.

Our next detailed document dealt with waste reduction, which needed to build in the basic mechanisms of keeping records, tracking, monitoring and evaluation of the effectiveness of the program:

Waste Reduction Program:
- Develop waste management contracts which provide for a designated amount to be targeted for alternate recycling markets or for on-campus research.

- Develop and establish co-operative purchasing agreements (for recycled content products or environmentally sensitive services) which can be used by other New Jersey-based universities, colleges, community colleges. Search to see if school boards and government entities can also be included.
- Assist in the university-wide waste audit program (used to gather information needed to enhance waste management and waste reduction programs).
- Develop resource tracking and performance mechanisms and analysis reports to evaluate the implementation of these policies.
- Develop and implement a total 'Landfill Waste Ban Program'.
- Develop and implement the 'Rutgers Investing in Itself for a Sustainable Future Program' utilizing contract savings – from environmentally sensitive contracts, to fund additional on-campus research into additional beneficial waste 'elimination' contracts and programs.

The policy implementation strategy

You may have written wonderful documents, but can they be enforced? How can you ensure that the policies will be executed? We understood that this was an entirely new approach and that the whole community needed help with guidelines and step-by-step instructions. We created a strategy paper which set out those steps clearly so that every department could begin to achieve the environmental goals, gather evidence, monitor performance and evaluate the effectiveness of their program: it set out the plan and processes for the implementation of the new policy:

The Implementation Process:
Using the centralized Purchasing Division as an environmentally sensitive 'filter' ... to show that there is a direct opportunity to identify and environmentally sensitize *all* commodities entering the university system and to identify and target all commodities to the appropriate markets or waste management operations (using the total product 'life-cycle' system). In addition, it is ... clear that the role of the contractor is critical. Even more important, the assistance and involvement of *all* faculty, staff and students will provide purchasing and the entire university system with a sustainable program which will enhance the entire faculty, staff and student education and working experience.

Since the program will be incorporated through existing operations, funding should not be needed to commence the development and implementation process.

All policies that are evaluated through the seven-point evaluation process will be incorporated in the official operating procedures.

We then developed a format or set of elements by which each policy document could be evaluated and which proposed that each policy would be organized and addressed in the following format:

1 *Goal or objective:* write a goal/objective statement review and create a practical enforcement statement; when the policy is implemented, monitor/audit periodically record data; develop personnel impact statements.
2 *Site visits:* make every effort to visit sites involved.
3 *Financial data:* provide a financial analysis/report with a financial impact statement; when the policy is implemented, monitor/audit periodically and record financial data.
4 *Environmental impact:* show evidence that policy is contributing towards greater sustainability.
5 *Departmental presentation:* if a policy directly involves a department, this committee will schedule an appointment with the department chair/director.
6 *Promotion:* committee will promote and release policy information to the university community.
7 *Historical data:* accumulate data on each policy and utilize this information for educational presentations, reports and for publications.

There is common ground across all organizations

Rutgers University has proved that purchasing provides an excellent base from which environmental policies and laws can be implemented. In the following chapters, we will explore this concept further and offer clear examples and case studies on how you and your organization can participate in this process. Take a look at your environmental policies and laws and see if they can be blended into your purchasing process. You will be surprised how familiar some of the terminology will be and, once you make the connection between purchasing policy and environmental policy, you will begin the process of developing your environmental gatekeeping skills.

Rutgers University took the lead in building its value system into its contracts. Having an environmental gatekeeper in the purchasing operation enabled Rutgers to be responsible for protecting and enhancing the organization's contractual obligations while incorporating environmental preservation actions and sustainable development practices within the contract process.

Environmental gatekeeping is for everyone, but some of the hard questions have probably kept some of you from getting involved. You do not need special training or certifications to master this unique and important responsibility. Environmental gatekeepers are just individuals who want to make a positive environmental impact on the organization and the surrounding community, to preserve historical economic and quality assurance practices and goals, and to maintain the competitive business ethic while performing the task of professional purchasing.

As environmental gatekeepers, your role in protecting the values and environment of your organization is critical. What you do with this new found environmental power is the key to your organization's environmental performance and it could be the guide that you use to shape your own environmental contract management/environmental gatekeeping strategy.

3
Using Your Organizational Structure

In the previous chapter, plans for an environmental purchasing policy and a program to implement it and evaluate its effectiveness were outlined. This chapter emphasizes the need to work with the structure of your organization, the lines of power and command, and the channels of communication employed. These elements will enable you to embed your policy and program into the institution so that it will endure over time.

Identify the structure of your organization

The first step is to look closely at the hierarchies, committees and decision-making bodies of your institution. Once you have identified and clearly understood your organizational structure, you can embed your new environmental plan into the normal operating procedures. This will set the stage for a successful and sustainable environmental program. The plan must be developed and implemented into your organization as part of the routine working systems, using all the existing power structures.

In the late 1980s, environmental preservation initiatives did not exist at Rutgers University and there were no concrete plans on how to organize a university-wide environmental program. As we saw in the last chapter, it took several years to develop one to fit into all the existing systems of the institution.

Be aware of the context in which you are operating

During the late 1980s, something other than the environment was occupying all of our attentions ... the entire United States was suffering through a recession. At that time therefore, even if there had been an environmental plan, there was no individual or group with the power or resources to approach the university administration and request funding to establish, implement or enforce a program. The economic context of the time was critical: to stay within our financial means was paramount and of much higher priority than our plans to comply with all the environmental laws. Although there was a push for environmental action from behind the scenes, from myself along with a group of students and staff, this topic was not at the top of the agenda for the university at a time of economic anxiety.

One thing we did discover through this lack of resources: any development and enforcement of a university-wide environmental program would have to be attached to an existing department, and it would have to be a department with university-wide decision-making powers. Most importantly, it would have to be a department that was not going to be eliminated due to budget cutbacks resulting from the recession. Purchasing was identified as that department because we were essential to the running of the university.

It was proposed that the services of the purchasing department could provide the hub or centralized 'home' for the university environmental programs, and the director of our department agreed. Our conditions for this assignment were simple: the environmental programs in which we would be participating would have to comply with purchasing guidelines and procedures. The purchasing division interacted with every department on campus, and we were involved in daily decision-making. This department would seem to be the logical choice for most organizations, though other departments could serve in different institutions. The key thing is to make sure that decision-making powers lie with the division you choose, and that it has a mastery of organizational procedures and guidelines and, above all, a willingness to participate. As I was a member of the purchasing department, and one who had been pushing for these changes from the very beginning, our environmental program occurred sooner than in most organizations.

Now we are on environmental preservation 'Easy Street', right?

Not quite. It took four years to establish the environmental program at Rutgers University. Progress was slow in winning commitment and understanding from our executive administration. This was partially the fault of the enthusiasts – we were often pushing at the wrong door. The early message that we, the students and myself, were conveying was not being delivered through the university operational and decision-making system.

In addition, the early environmental purchasing plan could not stand up against the strict competitive contract guidelines.

From this important early phase, it became very clear that three things were essential to the success of our plans:

1 *Leadership support.* If the leadership of the organization does not go along with the program, it will not succeed, no matter how well the program is designed.
2 *Clear and universal communication of the program.* If your plan is not outlined and communicated to the entire organization properly, it will take for ever to implement.
3 *Perseverance.* Long-term persistence and patience are essential. These changes will not happen overnight. You have to design as many programs as it takes to find one that truly fits your organizational structure.

We tried many initiatives over four years, from 1988 through 1992, and the only success stories were the environmental policies, a few environmentally enhanced contracts, lots of letters to the university community and vendors, many student initiatives and a lot of great individuals with great ideas.

Finally, the break we needed ...

Although the early attempts to implement fully an environmental program at the university was difficult, we went on working, adapting Plan A through to Plan Z. Without being discouraged, we continued with persistent attempts to convince our university community and the executive leaders that there was a genuine need for environmental initiatives.

In 1992, the students from Students for Environmental Awareness (SEA) and I were finally successful in persuading the University Senate and president to adopt a university-wide environmental policy and to introduce the program to the university community. We achieved this by being very busy behind the scenes, educating ourselves and others as we went along. At times, SEA resembled a professional lobbying operation. There were letters to US senators, to the university campus newspaper and to the University Senate. I attended scheduled University Senate meetings and meetings with high-level university administrators. At the same time, I also attended Department of Environmental Protection hearings and meetings on environmental issues, provided lectures and interviews around New Jersey on what the purchasing process could do to enhance and promote environmental issues. All this vigorous activity was educating and training us all in the theory and practicalities of improving the environmental performance of a large organization.

Achievement No 1: getting the leaders on board

Finally, a letter from the university president, dated August 21, 1992, announced the new environmental program to the whole university community of faculty, staff and students. Extracts from this letter are given below:

To Members of the University Community:

Today, Rutgers, the State University of New Jersey operates one of the largest and most successful university-wide trash recycling programs in the United States. Rutgers' leadership role in the recycling of solid waste began in 1973 with a single student sponsored recycling collection center on Cook Campus. Since those humble beginnings, our recycling program has been expanded to all campuses with convenient collection points in each building. Our accomplishments in this important environmental area have been impressive. During calendar year 1991, more than 50 percent of our trash was recycled.

Governor Jim Florio has asked public institutions in New Jersey to attain recycling rates of 60 percent by 1995. I am convinced that we can easily exceed this goal. Moreover, Rutgers, The State University of New Jersey, has a responsibility to the citizens of New Jersey to develop and demonstrate innovative and practical ways to deal with the continuing worldwide problem of trash disposal. Reducing the amount of material which would otherwise become trash and developing markets for recycled materials are key focus areas for solutions.

The University Senate, by Resolution S-9201, is supporting a renewed and intensified source reduction and recycling effort. Along this line, they encouraged the formation of a recycling and source reduction committee.

Reporting to the senior vice-president and treasurer, the Recycling and Source Reduction Committee will be charged to:

1 Review and recommend practical recycling, source reduction and recycled products use measures
2 Recommend goals and objectives for recycling and source reduction
3 Encourage co-operative interaction between diverse members of the university community to instil a 'reduce, reuse and recycle' ethic
4 Recommend educational programs.

I have asked Vice-President for Undergraduate Education to include in this year's discussion of the undergraduate curriculum the Senate recommendation concerning a required core course in environmental citizenship.

The mandate in the letter is clear, but what shines through are the individual components which are consistent with the university decision-making process. Additionally, it was made clear in this letter that the new environmental committee was made up of diverse members of the entire university and that key administrators had been recruited to participate in the program.

Achievement No 2: communicating to the whole organization

This letter not only showed that the president was committed to the program, but it also reached all members of the university. Since August 12, 1992, the date of that letter, the environmental programs at Rutgers University have had tremendous success, and this is attributed to executive leadership and the willingness of dedicated individuals to support the executive decision-making concept. Part of this success and support mechanism came from the understanding of how decisions were made and how the environmental program could be added to that agenda.

Achievement No 3: persistence and perseverance in reaching your goal

The newly formed committee was now properly authorized to issue its first document which included its stated goal:

The Committee's goal will be the conversion of the policy options to policy practice without jeopardizing the integrity of the mission of Rutgers University. The committee will make sure that the programs will aid the university in complying with the established New Jersey State DEP and EPA laws and regulations.

Finding the answers to four key questions

The strategy used to make this happen has already been outlined in brief. The detailed working out of that strategy can best be shown through the answers to four vital questions to be asked of any organization. Finding the answers to those questions helped develop the environmental contract management program that is still used in Rutgers today.

Question 1: What is the administrative or operational structure of your organization and how are decisions made?

Rutgers, the State University of New Jersey, with over 60,000 students on campuses in Camden, Newark and New Brunswick, is one of the major state university systems in the nation. Making decisions within this huge system is a tremendous task. However, research, education, administrative and operational decisions are made every day and these decisions could have an impact on the organization's environmental program. We recognized that understanding how decisions are made at the university would help us begin to formulate our environmental program.

Rutgers University is governed by the president of the university, the board of governors, the University Senate, a series of vice-presidents and various senior level administrative and faculty personnel. The organizational decision-making process is contained within the University Regulations and Procedures Manual and, to give an idea of how complex this process is, I have extracted some of the opening paragraphs from the manual:

> University officers. The principal officers of administration for the university shall be the president; the university vice-president for Academic Affairs; the senior vice-president and treasurer; the senior vice-president for Program Development, Budgeting and Student Services; the senior vice-president for Academic Affairs; the vice-president for Research; the vice-president for Public Affairs and Development; the vice-president for University Administration and Personnel; the vice-president for Information Services and university librarian; the secretary of the university; and university counsel. The principal officer of administration for each of the university's three campuses shall be the Provost of each campus ...

What a list. It goes on to read:

> The president shall be the chief executive, academic, and administrative officer of the university as provided in the Charter and in the Bylaws of the board of governors and is clothed with corresponding authority. All assignments of duties to other officers in these regulations shall be subject to the president's interpretation and decision as shall the interpretation, within applicable law, of any regulation, policy, or practice of this university.

This manual continues in the same vein, and this section concludes sixteen paragraphs later. The point is that all key individuals know that change can only happen within the framework set down in the procedural manuals. With this in mind, the two environmental policies and the environmental contract management program were designed so that they would fit perfectly into the structure of the university. After having been presented to the University Senate and president, this policy and program were introduced by the president himself to the university community by means of the letter already shown earlier in this chapter.

Question 2: What can you learn from the example of other institutions?

It is fascinating to see what other institutions are doing by way of establishing environmental programs. During our program development, I began a long-distance friendship with a government official from Wellington, New Zealand. Ray Richards's document, *Positively Wellington*, for Wellington City Council, showed that similar programs developed as far away as New Zealand present striking resemblances to the Rutgers program. Some of the key points of the Wellington 1996–99 Interim Strategic Plan were the following:

> Wellington City Council's strategic direction can be summarized as follows:
> 1 We respond to the needs of the community
> 2 We concentrate on our core business as defined by our role in the city
> 3 We identify the city's weaknesses, and assist in building the city's strengths
> 4 We play our part in looking after the city's key assets
> 5 We promote the wise and sustainable use of the city's resources
> 6 We manage our assets and finances in line with our core business commitment and our long-term goals.

This example led us to consider what the established laws or regulations that govern the organization and dictate how it operates or functions were. Using this information, how can environmental preservation be filled into the overall scheme of operation?

Wellington City Council has incorporated its environmental program within the established laws and regulations it uses to govern the city. These rules and regulations assisted Wellington in the design of their environmental purchasing program, as shown in their Interim Strategic Plan (above). The full Strategic Plan contains the programs created from each of the six key points and the environmental policies and guidelines are complemented by the Strategic Plan.

Question 3: How does your organization interact and integrate with the surrounding communities?

While working with Ray Richards of Wellington City Council on their Green Purchasing Program, I realized that Wellington had placed a significant amount of emphasis on the social and economic future of the whole community. This led me to look more closely at the location of my institution. Rutgers University has three campuses located in three distinct areas of New Jersey. Just outside of the campus borders are communities with which we have to interact on a daily basis. However, in my own investigations in 1988, the university community was not involved with any significant environmental preservation activities in the communities just outside the campus. Since the university organization has a large economic and social presence in our surrounding communities, I believed that the environmental programs created at Rutgers could enhance and develop these communities tremendously if we included the community in our environmental plans, especially our environmentally sensitive product purchasing plans. Therefore, the current university environmental policy and programs contain environmental contract management initiatives which include the local communities in the execution of our plans. This policy of local economic development has become a cornerstone in our environmental programs.

It soon became clear that organizations all over the world were facing similar problems. While providing a four-day lecture program on paper recycling and paper residues (from the point of view of a purchasing official) for the Universidad Nacional de San Antonio Abad del Cusco, Cusco, Peru, I discovered that no paper-recycling program had ever existed in Cusco. A plan was needed, one which world be accepted and practised by the entire community, including Cusco. The lecture audience contained administrators from the University, faculty, students and most interesting of all, local merchants, business leaders and community leaders from the city of Cusco. The university's plan of action was twofold:

1 Commence a paper recycling program for the school, and
2 Involve the stakeholders from the Cusco community in the paper-recycling master plan.

This type of collaboration between a major organization (such as a university) and its local community is critical in designing a sustainable environmental program. It represents a good model for all of us to follow. The result of this lecture and of its description of a paper-recycling policy was the creation of the first paper-recycling program, one which involved the Universidad Nacional de San Antonio Abad del Cusco and the local city together.

Question 4: What process decides what the organization purchases?

This last question brings us back to the purchasing role. For most organizations, buying products and services is based on two factors: best quality and best price. But before we think about buying new products, we need to go back to a prior stage and look closely at the demand for products and services arising out of the daily activities of the organization.

Developing a partnership between the individuals who create the demand and the individuals who satisfy that demand, with a carefully negotiated contract, holds the keys for designing an environmentally sustainable system. It is obvious that purchasing departments can find 'greener' products for their organization to buy. However, it is much more challenging to go back to the users and ask them if they could use significantly less of that product, or even do without it altogether. The ultimate 'cost avoidance' strategy would be, of course, that they could do without it altogether. It is worth remembering that the basic list of possible ways to cut down on consumption has four elements: refrain, reduce, reuse, recycle. These four elements are listed in order of priority, according to how few demands they make on the earth. Once it becomes a priority to get our whole community to think carefully before they order something new, then great environmental opportunities (not to mention cost savings) can be created.

At Rutgers, we used various communications to promote this message. For example, the initial letter from the Recycling and Resource Reduction Committee of Rutgers to the university departments included the following:

We have been charged to:
1 review and recommend practical recycling, source reduction and recycled products use measures
2 Recommend goals and objectives for recycling and source reduction
3 Encourage co-operative interaction between diverse members of the university community to instil a 'reduce, reuse and recycle' ethic, and
4 Recommend education programs.

The Committee's goal, with your help, will be to phase in gradually these policies within your department.

Many of the goals of the program can be advanced through education of the students, staff and faculty.

Increased procurement of recycled content products can begin almost immediately. Working with the Procurement and Contracting staff, our current contract suppliers may be queried about providing you with competitively priced, equally compatible recycled content products. Additionally, we have developed a 'Recycled Product Guide' which contains a list of products and vendors.

We will document every step of the conversion of Rutgers to the fulfilment of the policies' goals, so that this university may serve as a positive example for universities and other institutions around the country. To that end, we will conduct a baseline survey of the procurement practices in your department, with follow-ups at six- and twelve-month intervals.

This is just one of the many letters that was used to communicate important information and create mutual understanding throughout the university. It was critical that the university community understood that the purchasing department was fully involved and working hard to understand the new climate. The purchasing environmental policies and my attendance at New Jersey State environmental hearings and meetings, chairing the university environmental committee, and lecturing to students on campus, all demonstrated involvement and eagerness to learn. You will have a rough time if you are simply dictating what others should do and are not fully participating yourself. If someone had a question or concern, they were able to call me and I would attempt to answer it and or address the issue personally.

Various techniques and styles were used to introduce and communicate the rationale behind the Rutgers environmental contract management program. The intention of these communications was to engage the university community into adopting and practising environmentally sensitive activities as a routine matter of doing business. In all cases, the goal was to instil confidence and was underpinned by the realization that behavioral changes are needed to sustain long-lasting practices and progress. The letters and communications used in this process were absolutely critical. If you plan to issue such letters, they should fully explain your intent and should also invite members of your organization to participate.

One example of our efforts to increase internal participation was an early memo we sent to all department chairpersons, deans and directors:

Re: Departmental Recycling/Environmental Awareness Liaison
In an effort to increase awareness and university-wide participation in
the Rutgers University Recycling Guidelines, Policies and Procedures,
the University's Recycling and Source Reduction Committee would like
a volunteer from each university department/division to become a
Recycling and Source Reduction Liaison.

A department liaison is needed so that each department has the latest information on the university's recycling, procurement of recycled
content products, source reduction procedures and the NJ state laws
which govern environmental policies.

Faculty, staff and student participation is crucial to the success of the
recycling and source reduction program at Rutgers.

Rutgers University has a responsibility to preserve the environment,
as has any large company. Environmentally conscious faculty, staff and
students have made the recycling program at Rutgers flourish.

By engaging the entire system (including the vendors), we were able to
catch the attention of multiple audiences by using only the most limited
resources (most of these letters were personally typed, copied, sorted and
mailed by me personally). Our vendors were also included in the letter-
writing campaign.

Putting it all together: one document to tell the whole story

We felt that it was necessary to take the environmental policy initiatives
one step further without losing sight of the overall organizational system.
To do this, we needed to develop a working document which brought the
entire university (students, faculty and staff) into the picture. In short, we
had to find the most effective way to engage and communicate with a
massive audience on a frequent basis. The most important document of all
was the one that outlined the newly created program for Environmental
Contract Management (ECM):

Environmental Contract Management
- *Definition*. The design, merging and institutionalization of environmental preservation, community social values and life-cycle analysis
  systems (including input-out systems) into the established procurement and contracting system, and the organizational-wide business
  and operation decision-making process. To design and implement
  programs which protect and sustain the organization's environmental stewardship, health, education/training and practices through

the power of contracts. ECM establishes a system which can identify all imported and exported commodities to and from the organization and creates and establishes policies and programs to increase the effectiveness of the organization's environmental preservation performance.

- *Rationale.* Utilizing the organization's diverse ecological, cultural, political and socio-economic community, environmental contract management will involve the research, design, and institutionalization of environmentally sustainable policies and programs through existing organizational administrative and operational functions. This will be accomplished through the establishment of environmental procurement policies and programs. The organization will, in turn, use these policies and programs to modify existing organizational contracts, operations, and educational/training programs using existing funding sources (through cost-avoidance strategies) and external investments.

Our environmental contract management goals were clearly stated in this document, as follows:

- To demonstrate practical applications and innovative technology which reflect the principle that appropriate technology should mimic natural processes
- To serve as an example to identify strategies to reduce obstacles and strengthen enhancing factors for large scale applications of 'real-life' practices and technology world-wide
- To further the goal of environmental literacy by integrating organizational policies and practices with sustainable technology by linking the organization's practices to established world-wide curriculums (or training), research, operations and partnerships. This will be done by on-site visitations, workshops, lectures, conferences, distant learning and continuing professional education programs, etc.

We included passages that explained our thinking about supporting the economic development of our local community:

After researching and designing the ECM program, the organization must negotiate with its community on how best to integrate the ECM program into the existing infrastructure, including regulatory structure.

We also envisaged the potential of this program to reach an audience world-wide:

> The program should be presented as a system which should involve stakeholders at all levels of government, educational, and business community. This approach will maximize interaction with a wide national and international audience. The ECM program should publish expanded documentation on how the ECM program will directly enrich curriculum, employee training, research, R & D, operations, and partnerships world-wide and establish ECM administrative programs at other institutions of higher learning, governments, health care institutions and businesses around the world.

This program was created for one particular university and was designed to fit the personality, goals and mission of our organization, including the purchasing department. To begin with, we conducted an evaluation and audit of all purchasing policies and procedures, then slowly we began to modify the existing contract language and specifications to meet environmental requirements. Environmental preservation initiatives and sustainability practices which were consistent with our environmental goals and yet which maintained all purchasing regulations and policies were gradually included.

Measuring our achievements

There are many ways of measuring achievements in environmental management and there are whole programs established to devise indicators. At Rutgers, we looked at two specific approaches: one which looked at our environmental impact as an institution and one which was based specifically on our purchasing activities.

Ecological footprinting

Mathis Wackernagel and William E. Rees's publication, *Our Ecological Footprint* (1996), argues that 'The ecological footprint measures human impact on nature. Because humans consume the products and services of nature, every one of us has an impact on our planet. This is not tragic as long as the human load stays within global carrying capacity. But does it?' They go on to say that the ecological footprint concept has been designed to answer this question and estimate people's impact. It does this by measuring how much in the way of natural resources people use today to sustain themselves:

Ecological footprint calculations are based on two simple facts: first, we can keep track of most of the resources we consume and many of the wastes we generate; second, most of these resource and waste flows can be converted to a corresponding biologically productive area. Thus, the ecological footprint of any defined population (from a single individual to a whole city or country) is the total area of ecologically productive land and water occupied exclusively to produce all the resources consumed and to assimilate all the wastes generated by that population, using prevailing technology. As people use resources from all over the world and affect faraway places with their wastes, footprints sum up these ecological areas wherever that land and water may be located on the planet. (Wackernagel and Rees 1996)

Although environmentally responsible purchasing is not explicitly mentioned, I immediately realized the powerful contribution that decisions made by purchasing could have on the ecological footprint of an institution. Calculating your organization's ecological footprint could give you a powerful metaphor to use in developing an effective environmentally sensitive purchasing program.

Criteria for success

Benchmark criteria enable a profession, business, industry or even a government to describe and measure its achievements over dozens of different categories and over time. Year on year, figures for each of the benchmark criteria will reveal the growth or decline, change and progress of that organization or profession. The categories or criteria which are selected to show successful achievement are highly significant and eventually do a great deal to shape the industry being measured – the tail wags the dog.

At Rutgers, we have looked at many sets of benchmarking criteria, including those developed and used by the Center for Advanced Purchasing Studies (CAPS):

CAPS Standard Cross-Industry Benchmarks:
1 Purchase dollars as a percent of sales dollars
2 Purchasing operating expense dollars as a percent of sales dollars
3 Cost to spend a dollar (purchasing operating expense dollars as a percent of purchase dollars)
4 Purchasing employees as a percent of company employees
5 Sales dollars per purchasing employee
6 Purchase dollars per purchasing employee

7 Purchase dollars per professional purchasing employee
8 Active suppliers per purchasing employee
9 Active suppliers per professional purchasing employee
10 Purchase dollars spent per active supplier
11 Purchasing operating expense dollars per active supplier
12 Change in number of active suppliers during the reporting period
13 Percent of purchase dollars spent with minority-owned suppliers
14 Percent of purchase dollars spent with women-owned suppliers
15 Percent of active suppliers accounting for 90 percent of purchase dollars
16 Purchase order cycle time (in days)
17 Percent of purchase transactions processed through Electronic Commerce
18 Percent of services purchases handled by the purchasing department
19 Percent of total purchases handled by the purchasing department
20 Average annual training hours per professional purchasing employee
21 Percent of purchase transactions processed via procurement card (CAPS Web site).

Though at first glance these points might not seem relevant to environmental preservation, just imagine the impact on our consumer behavior if specific criteria relating to recycling, source reduction, energy use, transport use, water use, etc. were included as a measure of the overall success of purchasing departments. Major efforts would be made by purchasing professionals to score highly on these measures. If the existing lists are read carefully and thoughtfully, this type of profile information can be useful in suggesting areas which could be extended to include environmental enhancements.

The CAPS criteria can be developed to reflect criteria of sustainability. For example, to take just one item from the list, item 13, 'Percent of purchase dollars spent with minority-owned suppliers', is a benchmark which could well document your interaction with minority-owned suppliers from your local community.

Summary and conclusion

Although Rutgers University is a specific institution (a land-grant research and teaching university), it functions and operates just like any business or organization with a mission: it has a political structure, personnel,

budget, buildings and operations, and an end product ... in our case, gradua-
tion of the student, the diploma and all the knowledge gained so that
individuals can contribute to the enhancement of society.

All the national and international environmental programs that I have
examined for this book were created by energetic people who used their
organizational procedures as a guide. They also operate and function with
similar characteristics to those of a university; they have a mission, they
have a positive end-product and this has led to the enhancement and success of their environmental programs. I have found that the success of an
environmentally sensitive program will be created out of leadership, innovation and the creative thinking process. The environmental preservation
skills that you need to work with are, in most cases, already within your
organization.

The key points anyone should consider when designing an environmental contract management program within their own organization are
summarized below although you will, of course, want to make adjustments
to fit your organizational profile:

- *Choose your headquarters.* Establish where the program is going to
  be administered from; pick a department with organization-wide
  decision-making ability. They must be easy to work with and they
  should have a person willing to work with you. Develop your working group team.
- *Get executive support.* Get executive leadership support which is docu-
  mented and charges you to execute your program.
- *Get purchasing on board.* Use your newly established ECM headquarters to commence discussions with your purchasing or business
  department.
- *Understand the decision-making process.* Study the decision-making
  processes of your organization and develop your environmental
  program using the existing mechanisms.
- *Learn environmental language.* Find out how things are procured and
  use the environmental language that is used in new laws and regulations to enhance or modify the existing procurement process.

Developing and implementing policies and programs to preserve the future
of our physical environment, while living and working in a world where
the rush to globalization is a priority, seems like an impossible mission.
Since my own private audit of environmental action, I have made it my
challenge in the last ten years to practice the initiatives that these environmental documents and organizations are preaching. It is my contribution

to make the act of preserving the environment part of the conscience of a large organization as we all live and work within this globalized tidal wave.

4
Creating Environmental Contracts

The process of obtaining specific goods and services has always been dictated by industry or by vendors. You ask for a product or a service and you may, or may not, get exactly what you want. The contract *does* dictate what you expect, but if the entire industrial field is not making what you need or want, the chances are you will never get it. Before I commenced my environmental gatekeeping activities, I always wondered who was in charge. Purchasing has the power, we write the contracts and we award them. But I viewed the environmental movement differently. For a brief period, I believed the competitive contract process was in jeopardy. Everyone wanted to be 'environmental' but no one wanted to pay the price, the increased cost ... why should we? We wanted to be environmentally as well as economically responsible and I believed a well-organized plan to break the financial deadlock of environmentalism versus least cost was needed. This was how the redesign of contracts began – to hit environmental targets at the same time as scoring on low-cost and high quality.

Basic purchasing tools

This chapter shows how to add environmental elements into the competitive contract process. But before doing that, we should ensure that we share a common understanding of the most important tools used by purchasing professionals. This will enable you to recognize the parallel instruments in your own country. The most frequently used purchasing or bidding items are the following (or their close equivalents called by other names):

- the Request for Information (the RFI)
- the Request For Quotation (the RFQ)
- the Request for Proposal (the RFP), and evaluation
- a Pre-Contract Questionnaire.

Request for Information (RFI)

This is a pre-contract form issued to gather information in preparation for the formal Request for Quotation (RFQ) or Request for Proposal (RFP) process. It is very rarely used, but it can offer the contract preparer valuable information which would otherwise be omitted from the formal bid process. This opportunity to gather extra information will be extremely useful if contracts are to contain new environmental criteria. For example, the supplier may have recycling, remanufacturing or packaging return systems operating for other clients. If so, and if you know this, it will be easier to stipulate these requirements in your new contract.

Request for Quotation (RFQ)

This is the process most frequently used as a prelude to formalizing a contract. Once specifications are written, these specifications are married up with the formal procurement and contracting documents to form the RFQ package. The RFQ is issued when you know everything about the commodity or service that you will be procuring and all bidders simultaneously submit their RFQ packages to you on a specific date at a specific time for a formal bid opening. The lowest qualified price submission is awarded the contract, provided of course that your organization has funds for that amount available.

Request for Proposal and Evaluation (RFP)

This process is used when not enough information is known about a commodity or service to formulate a comprehensive specification, but enough is known to issue a competitive bid which will be evaluated by predetermined criteria, by an RFP evaluation committee using a predetermined evaluation system. The formal procurement and contracting documents will be used, but the specifications will contain the itemized evaluation criteria. Each vendor will be submitting their RFP packages to comply with these criteria. The evaluation committee will evaluate and score each proposal received. They will then tabulate the scores received from members of the committee, and the contract will be awarded to the contract vendor with the highest (or best) evaluation score. I have always used the algorithm evaluation system for evaluating RFPs, because it is scientifically designed and I have found that it results in high quality contract awards. This system can be used to set up your predetermined evaluation method. A sample algorithm worksheet is shown below.

The Algorithm Method of Evaluation

1   The 'weight' is a measurement of the relative/critical importance to the mission of the organization requesting the proposals. Each category may vary, but the total must always be 100 percent.

2   The rating system is from 1 to 5, with 1 being the lowest and 5 the highest; each of the proposal criteria will be evaluated using the following rating system:

   1   little or no capability to offer what you need
   2   below minimum requirements
   3   meets the requirements
   4   exceeds requirements
   5   superior level of capability.

3   On the final tally sheet, each evaluator will have their own column to record their rating.

4   The numerical total is the sum of all evaluators' scores.

5   The weighted points are the numerical total divided by the number of evaluators times the weighted average in percent.

   For example: If one of the proposal categories is rated 30 percent, the number of evaluators is 3, and their scores are 4, 4 and 5, then the weighted points would be calculated as follows:

   weighted points = .3 (4+4+5/3) = 1.299

   The initial step should be to organize and rank each proposal category. When there is no advantage, the suppliers should be ranked the same.

Pre-Contract Questionnaire

This is very rarely used but it has proved to be quite valuable in, for example, the Rutgers University solid waste management contract negotiation process. This document asks each vendor a variety of operational, social and environmental questions. The answers given can be used with the successful vendor to negotiate additional enhancements to the contract. Such a questionnaire is therefore a very obvious document for you to use with your potential vendors, focusing their attention on the environmental implications of their product or service and on the environmental performance of their own businesses.

Ready to Go

The heart of the entire organizational environmental program lies in writing contracts using a life-cycle analysis of goods and services and yet staying strictly within the RFQ/RFP process. Life-cycle analysis considers the entire life of the goods or services under contract, including present and future implications of that contract for the environment. The appropriate environmental impact decisions *must* be made prior to releasing the document for the competitive bid process. However, prior to putting all this information into the RFQ/RFP procedure, a systematic process of gathering all the necessary information is required.

In most organizations, the purchasing department is strategically placed in a position where it can become the screen or filter for determining the environmental and physical health (the life cycle) of each commodity and the contract for it. The decisions that the purchasing department makes include what products come into your organization, how they will be used and for how long, when that product leaves the organization and where it will go – back into research, to recycling, to the manufacturer.

If you are able to get the most out of your vendors with a powerful contract in its standard format, why not raise the bar a bit higher and challenge the established purchasing process to demand even more? But can you do this for the same price and achieve all your environmental and social goals? We were confident that we could, and we were willing to take that chance because the purchasing system is a solid institution with strict procedures and guidelines. These were all used to develop the environmentally enhanced, competitive contracts that we use today at the university.

Ready, Set ...

As a purchaser, you are given the task of compiling the list of prospective vendors who could provide a product or service to you. You participate in the development of standard specifications – what will be provided and the terms and conditions to which the vendor must adhere. Next, legal documentation must be used to convey these specifications, terms and conditions. This creates your RFQ/RFP document. At Rutgers, we use the following guidelines:

- *The Competitive Bid Process*. The purchasing department will determine when competitive bids for goods or services are required by estimating the value, reviewing the specifications and delivery requirements.

  Establishing the competitive bid thresholds will determine the bidding guidelines: limited authority to users up to a certain amount may be authorized, up to a certain price threshold; written and/or phone quotes can be obtained to place an order, starting with a higher price threshold; the sealed competitive bid process may be mandatory, with public bid openings as a result.

- *Bid Specifications*. The purchasing department issues bid specifications in a manner that will permit fair and equitable consideration from qualified vendors. Specifications should be written as openly as possible to avoid the exclusion of potential bidders.

  The nature of certain goods and services may make it difficult to formulate specifications. In such cases, the user department may specify a brand name, model number, or item catalog number. The ordering department should include 'or equal' as part of item specifications. Any alternates offered by the low bidder as 'equal' should be reviewed for acceptability by the requisitioner and the purchasing department. If alternates are not acceptable, the requisitioner must give written justification for each alternate rejected.

  For public purchasing operations: this justification is retained in the bid files and shall be open to the public for review, if requested.
- *Bidding Procedures*. The purchasing department should maintain lists of qualified suppliers in most commodity areas and is responsible for selecting suppliers. User departments may also recommend suppliers. The purchasing department is the final authority to designate the vendor for any purchase. Pre-bid conferences or mandatory site-inspections may be conducted when it is apparent that the complexity of the request for goods or services warrant additional clarification to the suppliers.
- *Definition of a Responsible Bidder*. Demonstrates sound judgement and integrity; with a record of satisfactory performance and a financial status that will not present a risk to the organization in its contractual relations. If the director of purchasing (or higher purchasing authority) concludes on the basis of available evidence that a bidder lacks the ability to ensure adequate performance, their bid may be rejected, even though it is the lowest bid. The organization retains the right to reject bids when costs are over budget, when bids do not meet specifications or when it is in the best interest of the organization to do so.

As all these activities are happening simultaneously, the critical point to keep in mind is that this process must be fair and competitive. The competitive contract process is much more detailed and involved than this summary description, but the basic point is that this bid process should include terminology for *all* contracts which supports environmental sustainability. Reaching this goal has been the basis of environmental gatekeeping activities at Rutgers University. If environmentally enhanced contracts could be developed while following standard guidelines, then this would be a major accomplishment. This was important because, in 1988, some of my purchasing colleagues were concerned about the merging of environmental initiatives and the competitive contract, citing concerns over price increases and lack of co-operation from industry. With these attitudes held by colleagues, I realized that I was up for the challenge of my career.

Normally, what makes a contract competitive is that the product or service requested can be delivered or accomplished by multiple vendors equally. If you know that only one vendor can provide a particular product and they are the sole vendor with that capability, then you must revisit that contract and redevelop the specifications to open the competitive nature of the contract. In some extreme cases which include research, health and safety, this strict rule can be overruled and justified.

Since the competitive contract process is a serious legal issue and could potentially raise protest if not accomplished properly, it is necessary to do the homework very thoroughly indeed. Early in my purchasing career it became clear to me that industry itself has a lot of information that they would be willing to share in order to be part of your competitive bid process.

In 1988, after our experience with the 1987 New Jersey Recycling Act, I began to attend New Jersey Department of Environmental Protection meetings and conferences. Some prospective vendors were present at those meetings, with a variety of environmental products and services on offer. I did not see anything of interest for the university but I did see the potential for creating specifications and contracts which were similar to the 'dog-and-pony' shows that I did see. The idea emerged that if vendors were willing to design and develop products and services to *sell* to the various organizations to meet the strict environmental guidelines of that time, we could develop our own enhancements through our contract process. We could probably get whatever we wanted – industry would supply us with what they knew we were looking to buy. The challenge was to keep it progressive, but not *too* progressive; to keep the process competitive to find more than one vendor willing to make enhancements; and to put this entire process through our purchasing process.

... Go!

At the beginning, the focus was on the *source* of the products or services. Our concerns and aims were circulated in two letters in 1990, one to all prospective university vendors and one to all existing contract vendors, as follows:

Dear Valued Vendor,

Rutgers – the State University of New Jersey has encouraged a high level of environmental awareness to all its faculty, students and staff since 1987. We would like the vendors who do business with the university to contribute to these standards also.

You may already be contributing to these efforts by way of environmentally sensitive packaging, establishing return packaging programs, or utilizing recycled content materials or parts in your products or in the services you provide (including the re/de/manufactured process for various equipment and/or providing recycling services for multiple commodities).

To this end, we encourage and support all initiatives your firm has to share with Rutgers University in regards to environmental preservation and sustainability: something we could be doing in addition to our established environmental preservation programs.

Please write to me and let me know how your firm has benefited the university's policy and goal to reduce landfill waste/incineration, hazardous waste and/or increase recycling standards and technology (if you are a current contracted vendor) or how your firm could benefit or enhance the university's policies and goals (if you became a contracted vendor). This information will be shared with the faculty, staff and students.

In addition, we would like to maintain an environmental commitment/mission statement for your firm, in our purchasing files. If you already have this document on file, please send me a copy. If you do not have one, we would like you to develop one. Please contact me if you need assistance in the development of this document; I can provide sample environmental statements for your review and consideration.

Dear Valued Vendor,

Rutgers – the State University of New Jersey is aggressively seeking ways to reduce solid waste from our waste stream at the university. For the past four years, I have been involved with researching and implementing contracts to make Rutgers more environmentally sound. Many of these contracts have been implemented with the co-operation of our departments and vendors.

I am researching now a 'return of packing materials program' (return to your company). The departments which receive the bulk of your shipments (especially regularly scheduled shipments or large individual multi-volume shipments) could accumulate the packing container (in most cases, cardboard boxes) and the packaging materials in a clear

100 percent recycled content plastic bag. On a weekly or bi-weekly schedule (whichever produces the least amount of accumulation and space) as a shipment is being made to the university department, the cardboard containers and bagged shipping material would be picked up by your driver and returned to your firm for reuse.

The logistics of this type of program can be designed by your company to make the process adhere to the purchase requests of the university and not interrupt the intent of 'delivering the required goods or products'.

Your support in reducing solid waste is important. Please advise if you have a system which would satisfy our waste reduction goals or if you would consider the plan I have outlined.

If there would be any additional costs incurred by Rutgers University for instituting this program it should be outlined in your response.

These letters were developed so that we could find out what could be offered to the university and to commence the dialog between our purchasing department and the vendors. Existing contract vendors and prospective vendors have always had a wealth of information and when a contract (or possible contract) is involved, co-operation always seems to be extremely high. It was also my intent to use the information gathered in this process to justify future enhancements of the contract specifications. The responses to these letters were fascinating, so the window into what could be accomplished was now being opened. For example:

Re: Packaging and Shipping Materials; Reuse and Reduction Plan
To assist you in achieving your recycling goal, we will gladly co-operate and reuse our packaging materials.

We indicate the following locations as delivery points to your facility with any regularity.

(Listings given of four preferred locations, plus frequency of collection and day of week of collection offered).

Our truck driver will pick up the packing material as indicated above ...

Please contact me when you are ready to begin the program or if you have any questions.

This program is at no additional cost to Rutgers University.

We thank you for your business and look forward to assisting you in this new endeavor.

(Signed).......................Operations Manager

The strategy was to have the vendors participate in the goals of the university and not drag them into our program, kicking and screaming.

In addition to these letters, round-table discussions were held with vendors and other purchasing officials and government agencies about their interpretation of environmental laws and – most important of all – what they were doing to comply. Not very much information came from government officials, and the other purchasing officials were just getting started, so I was flying solo at this point. The most valuable gain from this initial process was the gathering of basic information to support the competitive contract process. By using these letters, I opened the door for dialog and welcomed in innovation and fresh ideas while obtaining information that I could not obtain anywhere else. This information-gathering process also falls within the standard contract investigation criteria and thereby meets the competitive bid criteria. These letters are not unique, they can be utilized by most organizations universally. The task is simple and 'if you don't ask, you'll never find out'.

If you are concerned that there will be an added cost of environmentally enhancing your process, be reassured that this need not be so. Information is sent out to vendors all the time and if they want to be added to the vendor listing or just want to discuss the purchasing process, a copy of the environmental participation letter is always included. From the start, vendors must participate in the process if it is going to be successful. Once responses to the letter start to come in, you will see how willing both actual and potential vendors are to satisfy the environmental and financial aspects of your contract *in order to get your business*. It might be possible to create a program without going through the letter-writing campaign, but why take that chance? The more you know before you start, the more time you will save yourself in the end.

What Next?

Shortly after these letters went out I began to focus on 'total cost of ownership' (TCO) analysis and life-cycle analysis issues, and their relation to the competitive contract process. Basically, I was curious about how products were made, how long we could use them for, whether someone else could use them when we were done with them, and when we were done with them, how we could dispose of them: in short, what the total value of that contract to my organization is. I wanted to see everything upfront and right now!

To illustrate the TCO concept, I borrowed some interesting information from research conducted by Dr Lisa M. Ellram CPM who has also developed a Total Cost Modelling in Purchasing Focus Study for the Center for Advanced Purchasing Studies.

The first question we need to ask must be: 'What is Total Cost of Ownership Analysis?', and Dr Ellram proposes the following answer:

> There is a range of activity that fits into the classification of TCO analysis. In brief, TCO analysis is a structured approach for understanding the total cost associated with the acquisition and use of a given item or service. This may be compared with similar analysis from other suppliers, changes over time, or costs of producing internally versus outsourcing.
>
> An organization may use TCO analysis for one or all of the above types of analyses ... Based on the data collected from case studies of eleven North American organizations engaged in some type of total cost of ownership (TCO) modelling in purchasing, the researcher concludes that TCO analysis can be an extremely beneficial approach to purchasing. Total cost modelling, or total cost of ownership, means different things to different people ... The range of analysis that fits into the realm of understanding TCO in purchasing runs the gamut from operational, as in understanding supplier costs to aid in supplier selection, to strategic, as in re-engineering processes and supporting major outsourcing decisions. All of the organizations studied believed that TCO analysis was a worthwhile activity, and expected it to continue at the same level, or increase in frequency and importance within their organizations. (Ellram 1994)

Keep in mind the competitive contract

While developing purchasing functions and contracts using the foundations of TCO, I began to focus on what the true value of modifying contracts with environmental enhancements could be to our organization. With our new-found knowledge, we wanted to put all of these elements into the initial contract. However, we had to keep in mind that all vendors must first be given an insight into what this type of contract-writing was trying to accomplish. The simple and straightforward days of contract vendors delivering a product or service, submitting an invoice, getting paid and leaving the campus, were almost over. Our philosophy was: 'Here is our environmentally and socially enhanced contract and we want you to stay a while, get to know and understand us, enjoy the campus and discover what we can do together. Then we can discuss additional mutually advantageous initiatives'. We believed that a contract could be designed which would engage a vendor for more than the basic financial transaction and the competitive contract would be my weapon of choice.

The competitive contract must not be compromised and, therefore, you must look at the total purchasing process and stick with the purchasing basics. Your job as a purchasing professional should not be abandoned. However, it is imperative for you to leave your desk and go out into the field to investigate what is being done and how this information can be used to enhance your existing process.

Look at the big picture

At first at Rutgers, we always concentrated on the immediate results of the environmentally enhanced contracts. Though these were satisfying and brought economic benefit, it was not until we began to look at the full cost-accounting and life-cycle analysis issues as they related to the competitive contract that it become possible to see the full value of what we were do-ing for the organization and, interestingly, the full value of what these contracts were doing for the vendors and suppliers.

Vendors and the departments you serve must know your position on environmentally responsible purchasing issues and you must take con-trol. For example, vendors can dictate to you what they are willing to do but you have the power of the contract on your side. The contract must be clear and it must dictate what your organization values most. In addition, the departments you purchase for will not participate if they do not know what you are doing and how they can participate. If they know that you are looking out for the organization's best interest and you are utilizing the power of the contract to accomplish this, your communications and success with dealing with each department will be quite rewarding. With the investigatory information that you obtain and your mastery of the contract process, you will be well on your way to designing environmen-tally enhanced contracts.

5

Contracts in Action: Waste Management, Recycling Paper and Other Commodities

Writing contracts has always been an adventure for me. Once you start developing environmentally enhanced contracts, you may agree with me that the standard contract will never be as boring as it used to be. The environmentally enhanced contract brings new responsibilities and it can fully represent the best interests of your organization. The power of the contract can and should be used to complement and implement environmental policies, to introduce environmental educational programs and to develop sustainable environmental performance. This has rarely if ever been done. But by failing to do this, your contracts fail to represent your true social and environmental values. If you believe that the environmentalists and the politicians together will take sole responsibility for the future well-being of the planet, then you must have a lot of trust in the current system. However, as an individual with the power to negotiate contracts for a variety of goods and services from around the world, you also possess the power to make significant changes, power that you probably did not realize you had. And if you achieve this new form of contract management, your position within the organization will be well respected.

Purchasing policies, procedures and guidelines have always been value-based. The act of purchasing follows the same ethical code which protects the organization from economic hardship while keeping the purchasing process honest and competitive – in short, keeping purchasing officials out of jail. This process, if executed properly, enhances the value system of the organization. The contract is therefore a 'value implementation tool' and the environmental preservation enhancements that can be made will also be competitive, cost-effective and represent what your organization values most. Writing environmentally sensitive contracts will be your main environmental gatekeeping duty and your primary weapon in improving environmental performance.

It is helpful to set down precisely what are the goals for this new form of contract. These goals may be formulated as follows:

> Environmental Contract Management (ECM) involves the merging of environmental preservation, sustainable development, community social values and life-cycle analysis criteria into the established procurement and contracting process and specification. These contracts should be used to assist the organization in implementing programs which protect and sustain the environmental stewardship, health, education/training and practices of the organization. The contract should be used to identify all imported and exported commodities to and from the organization and the contract should contain criteria to increase the effectiveness of the organization's environmental preservation performance. (Rutgers 1992)

Making Environmental Contract Management (ECM) fit the system

Historically, there has been no requirement to screen contracts for their environmental provision. The purchasing official thinks it is the responsibility of the user to include these elements, and the user believes it is the purchasing official who should be aware of the criteria for environmental responsibility. In reality it is the responsibility of both parties, but the purchasing professional has the most significant burden. In most cases, he is the last one to see that contract document before it is released to the vendors for competitive bidding. At this critical stage, the purchaser must have decision-making ability to enhance the contract specifications and the contract language without jeopardizing the intent of the contract or artificially inflating the cost or value of the contract. If you visualize the complete cycle of the contract process and take ownership of the contract from design to grave, you will begin to realize what needs to be done.

Modifying the process

The standard purchasing flow process can be outlined as follows:

- Demand for goods and services
- Specifications are designed

- Purchasing department is contacted
- Purchasing merges specifications with purchasing documents
- The competitive bid process commences
- The bids are returned to purchasing for evaluation
- The contract is awarded to the qualified vendor
- Delivery/shipping/execution of goods/services
- Payment is made (after acceptance of product/service)
- Product usage/services provided
- Disposal of product/services terminated.

Now let us take a look at each of these steps and list the tasks associated with that particular stage, seeing how environmental considerations slot in alongside.

Demand for goods or services

- Compliance with organizational environmental policies is checked.
- Communication with the purchasing department during the creation of the demand. Purchasing then consults with the user's department about environmental purchasing requirements.
- Pre-contract questions are revealed such as how long will products/services be in use? What are the energy requirements? Are there any environmentally enhanced equivalent products? Can services be performed using an environmental preservation/sustainable ethic?

Specifications are designed

- The user department works with the purchasing department on the development of environmental performance and life-cycle criteria for vendors to follow, for example, manufacturing, product design, energy efficiency, sustainability, recycled content, packaging and disposal issues are considered during the process of designing the specification. All the examples in this chapter demonstrate how these elements were included in new specifications, achieving dramatic savings in costs for our organization.

Purchasing department is contacted

- Since purchasing has been part of the entire process from the beginning, this stage is reserved for additional research and investigation by the purchasing department such as surveying the market for additional vendors, interviewing vendors, reviewing price trends for environmentally enhanced products.

Purchasing merges specifications with purchasing documents

- Once again, this step is automatic since the purchasing department and user department have been collaborating from the start. This step should give the purchasing professional further opportunities to review the environmental purchasing policies for compliance.

The competitive bid process commences

- The terms and conditions of the contract are reviewed and enhanced to include contractor environmental responsibilities and for the development of contract vendor environmental partnerships.

The bids are returned to purchasing for evaluation

- The purchasing department checks vendor responses for compliance with environmental criteria, product design criteria and price and notifies the user department.

The contract is awarded to the qualified vendor

- The contract vendor is reminded of environmental responsibilities and a timetable or outline is established for environmental performance criteria for the contract vendor to respond to. In most cases, the contract vendor will already be responding to the purchasing department for quality assurance purposes.

Delivery/shipping/execution of goods/services

- The user department will work with the purchasing department to ensure complete compliance with all environmental initiatives established in the contract specifications. The vendor will submit to the purchasing department the reports necessary on all environmental criteria required in the contract specification.

Payment is made

- Payment is executed once compliance with all criteria of contract specification is made which includes all environmental performance criteria.

Product usage/services provided

- Throughout the life of the contract, the user department and the purchasing department will monitor the contract performance. Environmentally sensitive modifications to future contracts will be based on the data obtained from the performance of all environmentally enhanced contracts.

Disposal of product/services terminated

- The contract specifications should have had provisions for dealing with disposal, recycling and return-to-manufacturer issues. The contract vendor is contacted and all disposal issues are carried out in accordance with the specifications. Some of the criteria contained in this part of the contract are packaging return, reprocessing or re- or de-manufacturing of products by the contract vendor, exchanging of used products to another user department within the organization, composting at the organization's own site.

The environmentally sensitive purchasing flow process

The environmentally sensitive purchasing process is developed by *adding in* the necessary modifications at the stages shown on the table, on the right hand side. Once the system has been outlined as a series of clear steps, you can begin to create the environmental enhancements in the specification language (See Table 1 opposite).

Putting it all together: creating contracts for waste management

One important indicator of the responsibility that an organization takes for its own survival, and for the communities that surround it, is the amount and type of waste it disposes both within and outside its borders. It has been a historical belief that when we throw something away, it disappears magically and it becomes a problem for someone else to deal with. As a purchasing professional, you can execute contracts to identify and greatly reduce or eliminate the environmental burdens that your organization puts on society. In most cases this can be done with a significant cost saving.

When environmentally enhanced contract design began at Rutgers, new waste management and recycling contracts were being written at the same time. You may not be responsible for waste management or recycling contracts in your organization, but such contracts contain language which can be extraordinarily useful as a guide to the development of almost all environmentally enhanced contracts. All organizations have to deal with waste every day in one way or another so no one is totally exempt from participating in the proper execution of an effective waste management and recycling system. The information gained from waste management contracts can be used to query individuals who are responsible for other contract bids. Just asking the right questions about waste and disposal is a useful way of getting started with all environmental contracts.

Table 1.

Original system	Environmentally sensitive purchasing flow process
	Organization's executive, president or CEO provides leadership
	Organizational environmental policies created for the entire organization
	Environmental purchasing policies created that are specific to purchasing
Demand for goods or services	
Specifications are designed	Environmental policy enhancements are incorporated
Purchasing department is contacted	
Purchasing merges specifications with purchasing documents	Environmental policies of both the organization and the purchasing department are screened and merged with entire contract document
The competitive bid process commences	
The bids are returned to the purchasing department for evaluation	
The contract is awarded to the qualified vendor	
Delivery/shipping/execution of goods or services	
Payment is made after acceptance of product/services	Certain environmental criteria may need to be met before final payment is released
Product/services provided	Evaluation of performance requested by the user department
Possible disposal of product/services terminated	Recycling, composting, remanufacturing, return for further research on life-cycle or recycling technology, internally or to contract vendor. This information is contained in the original contract specifications

Those in purchasing departments who are used to waste management might like to take the reverse approach. First take a critical look at all the other contracts and evaluate what these contracts contribute to the waste management picture. If you look at this emerging monster, you will begin to see that waste management contracts alone do not protect the organization from infringing environmental principles but they do provide a useful structure which can be used in the design of a great number and variety of other contracts.

Take a simple example. If you have a contract for the delivery of furniture, the amount of packaging that comes with chairs, desks and tables can be enormous. By negotiating return packaging, or by asking for the furniture to be delivered wrapped in reusable blankets, both the environmental and the financial burden of your waste management contract is reduced.

Now consider all the contracts your organization executes. By concentrating on packaging alone, the possible financial reductions will be considerable.

As stated earlier, recycling became law in New Jersey in 1987. The goal was to design waste management and recycling contracts which ultimately reduced the amount of waste that was being landfilled or incinerated. For Rutgers University, if contracts were executed properly and compliance achieved, this was an opportunity of a lifetime. Since landfilling and incineration were more costly than recycling, it made sense to focus on the economic advantages of executing a waste management and recycling contract which shifted the focus towards increased recycling and waste avoidance. Therefore, the first step was to look at all the university's contracts and to see what contribution each contract was making to wider waste management.

The waste management contract that Rutgers issued in 1990 was executed as a RFP (Request for Proposal). The contract specifications were prepared with the co-operation of the Facilities Maintenance Division, Purchasing, Housing and Dining Services. The specifications contained criteria which could assist the university in reducing its waste burden and each criterion was assigned a numerical value. When the contract was issued to the vendors for competitive bidding, each was instructed to respond to our criteria by explaining how they would execute and provide the services we valued. When the proposals were returned to purchasing, a committee of individuals from Facilities, Purchasing, Housing and Dining evaluated the responses given by each vendor using a mathematical algorithm schematic. Finally, a contract award was made. This was an effective contract because the reward for reducing solid waste was one of the most significant features in it. In addition, a separate contract was written for the landfill. This was done for a unique and very specific reason. The contracted waste hauler was given the contract to collect our waste and deliver this to the landfill or to take our recyclables to market. The contract to the landfill was executed to pay for any waste that our contractor delivered to them. Therefore, if we reduced our landfill waste stream by a significant amount, we could reduce the financial burden on the organization as well. As a result, our contract savings increased with our increased recycling rate. From 1990 to 1996 the university saved over $1.2 million in waste management costs.

Developing a fully functioning waste management and recycling contract is very time-consuming and it requires extensive self-examination within the organization. However, outlined below are some of the key elements that you and your colleagues will need to consider in developing and designing waste management and recycling contracts which reflect the values of your organization. This information is derived from all the waste management and recycling contracts that I have written and researched since 1990. However, the contract information in the lists below is a personal view of what should be included in the contract design process and you may wish to alter, delete, modify or enhance any of this information to fit your own needs. For simplicity, the package is based on the RFQ (Request for Quotation) process. An RFP may be used, but you will have to design the evaluation criteria and develop your evaluation committee (all prior to releasing the RFP package for competitive bidding).

Key elements for a waste management strategy

The key elements are, in summary:

- *Cost*. The pre-emptive strike – hit before you get hit.
- *Quality*. What do we get and how can we participate in the plan.
- *Environmental*. Protecting our values; establishing the standards and measurements.
- *Economic*. What our contract can do for our economy; investing in ourselves.
- *Institutional and legal*. Mandating and outlining the value system; delivering the plan.

Cost. The pre-emptive strike – hit before you get hit

Take measures to enhance cost avoidance, such as source reduction and diverting waste away from the waste stream (and from landfill), before they become a cost consideration. Examples of approaches include the following:

- Separate contracts (hauler versus waste receiver)
- Package return/product return contracts
- Writing contracts for commodities which have a waste management clause written in them

- Cost strategies that reflect and complement the entire contract theme (i.e. cost avoidance/waste reduction)
- Cost categories (i.e. cost per home/industry/ton/volume/container size/ type, type of service, commodity)
- Energy savings, infrastructure savings (hidden cost savings, less truck traffic on campus)
- Consumers: bundled cost (usually buried in taxes)
- Revenue sharing plan (percentage of gross receipts returned)
- Recycling rate guarantee program (rebate of x percent per ton for every recycled percentage below target recycling rate (for resource reduction goals)
- All other costs incurred (direct and indirect).

In addition, real cost factors should be reflected in your pricing strategy.

Quality. What we get and how we can participate in the plan

Design a system which closely matches existing capabilities and the mission of the organization and add incremental enhancements with education and training and upgradeable technology. Contract responses should describe and outline their entire proposed system. Examples include the following:

- Logistics/plan of operation, including size and scope, difficulty in behavioral changes of community (how do we get our waste from A to B and, in the future, remanufactured waste from B to A?)
- What is the frequency of your plan (flexibility with volume increase/ decrease in services)?
- How far-reaching is the accepted strategy prepared to be?
- How much is known about our waste commodities and what will be done with each (considering cost, cost avoidance/reduction plan, quality, environmental, economics, institutional/legal)?
- Where has the technology/contractor positioned themselves, industry-wide (who is doing what, where)?
- Past experience of the contractor; references from other clients.

Environmental. Protecting our values; establishing the standards and measurements

Examples include the following:

- Comprehensive waste audit/assessment (before, during, after contract period), i.e. quarterly reports due (frequency to be determined in contract)

- Risk assessment reports
- Recycling rate or environmental goal (should be numerical to be met each year)
- Measurements of environmental standards (comparisons) against similar institutions (frequency of reports to be determined).

Economic. What our contract can do for our economy; investing in ourselves

Institutionalize the life-cycle analysis strategy/program. Examples include the following:

- Involvement of local community and its values and concerns in the contract
- Use closed-loop systems (collecting waste and also buying back the recycled goods made from it) in order to involve participation of community (and create jobs)
- Use cost avoidance savings to invest in the development of technology transfer programs (contract allows input from you to contractor)
- Find new sustainable market strategies as a proposal request
- Do-it-yourself facilities using your own technology developments (full service recovery and processing facilities).

Institutional and legal. Mandating and outlining the value system; delivering the plan

The policies, regulations and legislation that govern your entire program and contract. Examples include the following:

- Public awareness program (outlined by contract specifications, executed by contract vendor/s; informing customers of continual support which provides valuable information exchange)
- Negotiation clause (open-ended enhancements/alterations)
- Length of contract (extensions), termination clause
- Flexibility necessary owing to law changes
- Vendor/customer partnerships
- Citations received by contractors: pre-contract disclosure information.

Developing a checklist

Before you start to design your waste management and recycling contract, spend some time creating your own checklist. Use your list to help educate your colleagues in advance, prior to creating a contract specification

which should contain all the criteria that you value most. Your preliminary checklist is likely to include the following points:

- Relate to your organization's environmental policy/sustainable develop-
  ment program
- History (current program, waste generated)
- Applicable laws, regulations, population, geographical area, site selec-
  tion(s)
- Environmental policies and implementation/evaluation plan (local,
  city, county, state)
- Environmental education/training requirements (for your staff, customers, community)
- Budget, current, history (distribution of costs), revenues collected, stran-
  ded investments
- Insurance and other liability regulations
- Key players involved in the design of contract (such as mayor, commu-
  nity authority, freeholders, business, state, federal, educators, school
  district administrators, administrator, main contact)
- Economic development opportunities (list), job creation and sustainability
  (associated with environmental management – waste management),
  jobs created – where/how, how many (real jobs)
- Communications (public relations, marketing, advertisements)
- Site visitations (community, markets, industry)
- Risk assessment (studies, data, information, insurance)
- Environmental impact studies
- Pollution prevention assessments
- Enforcement (compliance) responsibility/authority
- Reporting (performance); data collection; waste audits
- Public/community involvement (community volunteer programs)
- Partnerships (joining forces with government, other communities, etc.)
- New technologies (what do you know, university research, technology
  transfer opportunities)
- Develop the decision-makers' flow-chart (from 'concept' to 'action/prac-
  tice').

Contracts for paper with recycled content

It is always particularly difficult to convince an organization of the value
of a recycled paper contract. Recycled content paper for copying and for
printers (laser, desktop, high production) has always been more costly and
continues to be more expensive than virgin. There are many reasons for

this but, in my investigations, I found out that the most significant reason for the price difference was in the cost of purchasing high quality recycled content paper to add to the mix.

In short, I found out that the paper mills which were supplying paper to Rutgers had to purchase high quality scrap paper on the open market and add it to their paper-making process (post-consumer waste). In addition, the amount of investment required in machinery that produces recycled content paper is just as significant as in the equipment purchased to make virgin paper. As a result, if there is an increased demand for recycled paper, the paper mills will need to invest in new and improved equipment. This will be costly and this added cost will be passed on to the consumer or purchaser in high recycled paper prices. Several early attempts were made to write contracts for recycled paper for the university, but virgin paper was always cheaper.

In 1993 I began the research and investigation into developing a system by which Rutgers would dispose of its white paper waste directly to the paper mills which produced our paper in the hope of driving the cost down to more favorable levels; this program was called the R-Plan. The final chapter in this book describes the student involvement in this project in more detail, but the concept was unique and the responses we received were quite favorable. The contract documents prepared for this program were not awarded at the time, although it is planned to use them in future. Even so, the process enabled the university to understand the market forces in recycled paper much more clearly and to enhance the current RFQ so that it could be used with suppliers.

The first step was to issue a memo to prospective suppliers explaining our quest for low-cost recycled paper. Two fundamental elements included were, firstly, the supply of Rutgers own waste bond paper to the paper mill and, secondly, the establishment of a co-operative purchasing agreement open to all New Jersey colleges, which would greatly increase the size of the guaranteed local market for the recycled paper. This memo read as follows:

Recycled Content Paper. Pre-Bid Request for Information
Rutgers – the State University of New Jersey, for the last three years, has been leading the state in recycling and source reduction initiatives. However, the procurement of recycled content bond, xerographic and laser paper continues to be the weakest part of our program. We are now investigating a new initiative which could increase the procurement of this product and increase source reduction and we want your assistance.

In 1991/2 I rewrote our paper specifications to increase the quality and performance of the recycled content paper stock we procure. Since then, I have written to our departments asking them to procure this more expensive, high quality, recycled content paper.

Rutgers University would like to establish a contract (through the competitive bidding process) with a paper mill or through a local paper distributor for the production of recycled content paper using our waste bond paper. Through a co-operative purchasing agreement, this contract will be open to all New Jersey colleges and universities to procure from.

Please state if your firm would be interested in this type of contract and any input you may have. Our intent is to procure recycled content paper at the same or lower rate than our virgin paper.

It was important to make clear to prospective suppliers that the university had the responsibility of supplying suitable elements of its own waste paper to the paper mill, making it unnecessary for the mill to buy in post-consumer waste to add to virgin pulp.

Meanwhile, it was necessary to convince the university faculty and staff that recycled paper was high in quality and worth their attention as part of the overall environmental policy of the organization. The information circulated to all members of the university read as follows:

The University Committee for Recycling and Source Reduction enthusiastically encourages the procurement of recycled content paper. The specifications for recycled content paper were written in order to provide Rutgers University with the *best* quality recycled content paper available at the lowest qualified price, through this competitive bid process. (The bid specifications required the recycled content paper to perform the same as virgin paper.) The procurement of recycled content paper and products is one way to protect the future of our environment and to help increase the recycling of paper waste.

The price of virgin pulp paper on this particular contract is at a record low, due primarily to the paper mills' greater production capacity with new paper-making machines that have recently come on line. Paper manufacturers contend that this is a temporary condition and virgin pulp paper prices will be moving upward significantly by the end of the year.

In spite of this temporary price difference between the recycled and virgin product, we should not be discouraged from buying at least some of the recycled paper. You may as a start, consider buying one (1) recycled carton out of every five (5) purchased. This kind of interest will cause the mills to increase recycled product production, making the price more competitive.

Please try the recycled content paper. The prices are constantly going down as the manufacturers see our demands increase; last year the recycled paper contract price was $22.99 per carton. This year the contract price dropped to $21.80.

Detailed information was then provided for prospective suppliers to ensure that they would deliver high quality recycled paper, as follows:

Recycled Content Plain Paper. General Specifications
- *Recycled content paper* means any paper having a total weight consisting of not less than 50 percent secondary waste paper material (30 percent pre-consumer, 20 percent post-consumer waste) within the total amount of paper delivered under the contract. In addition, priority consideration shall be given to recycled paper or paper products with the highest percentage of post-consumer waste material which can meet the performance specifications in the Technical Specification section.
- *Secondary waste paper* means paper waste generated after the completion of a paper-making process, such as:
  - post-consumer waste material.
  - envelope cuttings, bindery trimmings, printing waste cutting and other converting waste.
  - rolls and mill wrappers: except that secondary waste paper material shall not include fibrous waste generated during the manufacturing process, such as fibers recovered from waste water or trimmings of paper machine rolls, fibrous by-products of harvesting, extractive or woodcutting processes, or forest residue such as bark.
- *Post-consumer waste paper* means any paper product generated by a business or consumer which has served its intended end use, and which has been separated from solid waste for the purposes of collection, recycling and which does not include secondary waste paper material.
- *General specifications*
  - Paper must be no more than three (3) months old.
  - The grain of the paper must be parallel with 11 inch side of the sheet (long grained).
  - Must be free of all defects (holes, wrinkles, tears, turned-over corners and damaged edges) and any scraps or foreign material.

- There must be no wrapper glue on any sheets.
- All sheets must be square and to size.
- All pre-drilled paper (3 hole drilled) should be free of plugs and interlocking of sheets that are caused by dull drills or punches.
- The paper must be smooth.

This detailed information could then be used as a basis for the contract specifications. The full technical specification for the recycled content plain paper was very detailed but the headings for the different elements specified were as follows:

- Stock
- Absolute moisture
- Basis weight
- Stiffness
- Tearing strength
- Opacity
- Smoothness
- Curl
- Ruling and writing qualities
- Surface
- Size and trim
- Grain
- Color, formation, and cleanliness
- Performance.

Each bidder was also encouraged to enclose with their RFQ submission the following certified signed statement: 'The following mills manufacture paper using only an "alkaline chemistry" resulting in the product with a basic (alkaline) pH value,' and all bidders were required to list mills.

The logistics of this particular program are constantly being refined and should result in a contract for recycled content paper which is high in quality and lower in price in comparison to virgin paper. In the meantime, the information that was gathered during the R-Plan development process was used to strengthen the contract specification process which is currently employed. The current contract that we have is the best we could do using traditional RFQ guidelines and the best specifications. By December 1997 the cost differential was $24.61/case – 8.5 x 11 (virgin) as opposed to $26.60/case – 8.5 x 11 (recycled content 20 percent post-consumer content).

Other commodities

As stated earlier in this chapter, environmentally sensitive contracts have been introduced into many other areas, in order to support the university's new waste management and recycling contract. The first comprehensive waste management program provides a firm foundation on which many subsequent contracts will build, because they will be obliged to feed into it. Conversely, the various other contracts your organization issues may determine how successful the waste management and recycling contracts are. If they are not designed to line up with the waste management obligations, they can undermine your attempts to achieve your waste program objectives. All goods entering and leaving your organization must be covered by the same comprehensive waste and recycling program.

Some examples are given below to show how well some of these other contracts link into the waste program and what measures we have taken at Rutgers to execute them.

Food waste

Food waste is covered by separate contracts at Rutgers University. In accordance with New Jersey Department of Environmental Protection Regulations, we cannot dispose of food waste in our trash stream. A unique contract exists to satisfy this requirement. Equipment was installed in each dining facility on campus. Rutgers University wrote a contract to install these systems for the reduction of our food waste volume and also to prepare the waste for possible composting on campus.

Recycled content plastic garbage bags

This contract provided hidden financial advantages that were discovered during our pre-contract investigations. Financial savings resulting from environmentally enhancing contracts is always a winning argument with your organization, no matter how small or large. Since the virgin content plastic garbage bag is made from petroleum-based products, the price was fluctuating quite often during the late 1980s and early 1990s. By changing the contract specifications slightly, we discovered that the *clear* recycled content plastic bag was less expensive because much less petroleum is used in its manufacture. However, the contract specifications called for the same performance criteria, in this case, tear strength. Therefore, this contract modification yielded an immediate cost saving and the quality was not reduced. However, there was an interesting problem ... color. The prior contracts required many different color bags, each color representing a different size: for example, if you wanted a 32 gallon bag you could rely on the black bag. This was useful for some of the custodial staff who could not easily read English. To solve this problem, we negotiated with the contract vendor to color code the labels on the boxes containing the bags, using the same color as the bag which used to be in that size category.

In order to realize fully what we were getting, and to justify the cost comparison, we issued the contract with both the virgin and the recycled content bag in the specifications and we also provided a bid sheet for each category. We then circulated information to introduce the idea of the new bags to the university community as follows:

> Please discuss the possibility of changing the request for virgin-based color polyethylene garbage bags to clear recycled content polyethylene garbage bags with your superiors and the custodial staff. I have found that a saving could be passed on to Rutgers as a result of this change. The sizes, dimensions and specifications will remain the same.
>
> I am proposing that there be color-coded labels on the cases/boxes which will reflect the same color bag which presently corresponds to that size bag (i.e. the present requirement for the 32 gallon black bag would be changed to a clear recycled content plastic 32 gallon bag with a black label on the outside of the box).

Other contract modifications

By December 1997, the following commodities were being recycled at Rutgers University:

- Antifreeze
- Asphalt and concrete
- Auto batteries
- Auto tires
- Fluorescent tubes/ballast
- Contaminated soil
- Corrugated cardboard
- Food waste
- Glass
- Aluminum
- Plastic (bottles)
- Steel cans
- Leaves
- Laboratory chemicals
- Mixed paper

- Computer paper
- Motor oil
- Agricultural products
- Scrap metal
- Telephone books
- Furniture
- Appliances
- Toner cartridges
- Wood.

In most cases, the recyclability and recycling of these commodities was based on the initial contract being inclusive of recycling requirements. It is my hope that some of the items that feature in the recycled content commodities/products contained in the later list given in the Environmental Protection Agency's Comprehensive Procurement Guidelines (see below) were the result of early recycling efforts like the Rutgers University program.

Buying recycled

We have spoken before in this chapter, and earlier in the book, about the 'closed loop' of collecting waste and then buying recycled goods made from it. This means that the collection of waste for recycling and the supply of products made from recycled materials are processes that are interdependent and reinforce one another. Put another way, there must be a market for waste material once it has been separated out from the waste stream. That market depends on there being yet another market: one for the recycled or remanufactured goods that are created from that waste. The recycled 'closed loop' must function alongside – and in competition with – the production of goods made from virgin resources in the same economic climate. So real effort must go into supporting both sides of the equation: waste must be separated out and disposed of to industries that will use the waste for new products. These new products must be bought as a priority to ensure the 'closed loop'.

The US Federal Government was very well aware from the outset that the presence of the closed loop was essential to the success of its program of waste reduction and recycling. To this end, the Environmental Protection Agency has issued a list of products which can be purchased in the US and which are in accordance with their Comprehensive Procurement Guidelines, as follows:

- Paper and paper products
- Vehicular products
  - Engine coolants
  - Re-refined lubricating oils
  - Retread tires
- Construction products
  - Building insulation products
  - Carpet
  - Cement and concrete containing coal fly ash
  - Cement and concrete containing ground granulated blast furnace slag
  - Consolidated and reprocessed latex paints
  - Floor tiles
  - Laminated paperboard
  - Patio blocks
  - Shower and restroom dividers/partitions
  - Structural fiberboard
- Transportation products
  - Channelizers
  - Delineators
  - Flexible delineators
  - Parking stops
  - Traffic barricades
  - Traffic cones
- Park and recreation products
  - Plastic fencing
  - Playground surfaces
  - Running tracks
- Landscaping products
  - Garden and soaker hoses
  - Hydraulic mulch
  - Lawn and garden edging
  - Yard trimmings compost
- Non-paper office products
  - Binders
  - Office recycling containers
  - Office waste receptacles
  - Plastic desktop accessories
  - Plastic envelopes
  - Plastic trash gags
  - Printer ribbons

- Toner cartridges
- Miscellaneous products
  - Pallets

Environmentally enhanced contracts bring many benefits

The contracts described in this chapter can go towards meeting some or all the of four challenges to reduce consumption and waste in our profligate society: refrain, reduce, reuse, recycle. Careful thinking and planning to do without, to use less, to repair and reuse items instead of discarding them, can bring about the obvious saving of cost avoidance. That is your first priority. But if you must buy, think about the end destination of your goods and weave them into your closed loop by buying recycled.

Every opportunity should be taken to evaluate the environmental impact of the commodities for which you are writing contracts. Your contracts should include statements which mandate the recyclability or specify the recycled content make-up of the commodity. And, of course, the contract must be economically responsible.

In addition, you should decide whether one waste management and recycling contract vendor should handle your entire waste stream or whether your individual contracts should contain return for recycling or remanufacturing clauses. At Rutgers University, we had to use a combination of all of these scenarios. It was our determination that the contracting office could negotiate better market pricing for several commodities by including waste management and recycling provisions in several commodity contracts. The balance of the material was included in the campus-wide waste management and recycling contract.

6
Working Relationships

After the establishment of your environmentally sensitive programs, the system will need to be continually examined, audited, reported on and improved as time goes on. In order to keep the system thriving, it will be necessary to obtain a regular supply of fresh, energetic and forward-looking ideas. These ideas may come from within – champions can be found at any level and in any department – or they may be found outside in the wider world.

One way to sustain the future of environmental purchasing and contract design is to create strong links with, and to learn from, other members of the immediate, as well as the national and international community, who embrace the same fundamental and universal goals. The first part of this book stayed within one sector: purchasing and contract design for large organizations. This chapter looks at three different sources of new ideas, of fresh energy and drive, of interesting and unexpected viewpoints:

- within the institution (in a university, particularly by working with students)
- through international programs, and
- through the Internet.

It examines the benefits of working closely with others who can provide support and guidance. Such partnerships involve the free exchange and open sharing of expertise and experience that come from collaboration. The same sharing is not always possible in the context of business, where competitive advantage reigns supreme.

There are many groups around the world which are passionate about environmental improvement, are flexible and creative in their thinking about strategy and are dedicated and hard-working in implementing their programs. These groups spend 100 percent of their working lives thinking about the environment and networking and sharing with others. Naturally, with such intensity of focus, they collect a huge amount of knowledge and expertise which they share freely with all who need it. This chapter introduces some of those with whom I have worked and explains what I have shared with them and what I have learned in return.

During the last eight years I have been lucky enough to travel to many national and international destinations, conducting seminars and workshops, contributing papers to conferences, lecturing and sharing information on environmental contract management systems. Many of these institutions had some form of environmental management scheme within their existing plans, but were not vigorous in putting it into motion. One of the reasons for this lack of action was their failure to involve their purchasing or business departments in their planning or decision-making process. However, the partnerships that developed at these various locations allowed me to share the experiences that I have had in New Jersey, USA, and to encourage them to go forward along the lines this book describes. Equally, I was able to take back with me the best information that they had to offer in their environmental program design.

This merging and sharing of information has proved to be extremely successful in all the partnerships with which I have been involved. I am certainly no expert on community or social science or on any of the many technical environmental programs but, by now, I know where to find this information and this is all I need. The contacts that I am highlighting in this chapter have really helped me and have often become personal friends. These are important relationships to anyone who is trying to change their institution from the inside. No one can do it alone.

Within the institution

First of all, let's remain within our own institution and look at what needs to be done. In order to sustain the environmental management programs in any organization, an established and ongoing environmental literacy and education program has to be instituted for the whole internal community of that organization. In a university, that will include faculty, staff and students: elsewhere it will include all staff as well as all users of services.

After identifying your organization as the 'real-life' working laboratory for positive environmental change, the future for the program will rest on how the organization integrates these programs and policies within its operations and management, its main activities (which in a university are teaching and learning), and within the staff training experience. Environmental literacy training is the technique used to train individuals in how to integrate environmental issues into many different levels and settings. Then, and most important, comes the task of institutionalizing – embedding – these changes into the organization's systems, finding ways in which this approach can be established within each area of expertise on a permanent basis.

Working with students

In a university, the student body represents a large and powerful force which can be harnessed and this group provides the focus for this section. In other institutions, staff, employees, users and consumers, and the groups representing them, will be able to play a similar role.

Students have many diverse backgrounds and interests, and there are often a great number and variety of environmental educational programs at universities and colleges. Students world-wide have contributed a great deal to the improvement of the environmental performance of their universities and one can learn a lot from them. Young people may have learned a lot about the environment while they were at school, from enlightened teachers, or from their families, or from the media, and they bring this knowledge and commitment with them to college. Attend an environmental meeting on campus with faculty and staff and you may come away knowing a good deal more than you did when you went in. If the students are a particularly vigorous and dedicated bunch, you may come out very impressed indeed and thinking that you are way behind in your learning curve and that they are at the forefront.

There is another reason why students have something to teach us: historically, they have frequently challenged the norm – let's hope this never stops. By challenging the accepted style of management, they are asking us to justify the rationale behind the decisions we make. This challenge is usually positive: students do not want to see administrators and faculty fail, they want them to succeed. Moreover, students want to be part of the solution, even though they may have a different approach to the common goals we all share. Above all, they are independent and free at this point in their lives. They have not put on suits and signed on with a company, afraid to put their head above the parapet in case it hinders their career progression.

Knowing that the past and current generation has misused and overused the earth's resources, our students are trying to lay claim to what is rightfully their future. They are on a mission to clean up the mistakes of the past and to slow down the deterioration process while setting a new plan for a sustainable future. The lessons we can learn from some of these visionary students will help shape the new vision for future generations. Below we suggest two ways in which this could be done:

- *An induction course on environmental performance*
  New students – and new employees – are usually given an introduction to the mission, the goals and the operational systems of the organization they have just joined. This is an ideal time to catch their interest in the organization's environmental performance and ensure their support for the future.
- *Putting the environment into the core curriculum*
  As students or new employees enter the school system or industry to receive an education and training for a profession, they must have (at least) an introductory understanding and usage of basic language, social and math skills which will carry them throughout their academic or professional career. In the future, might there not be included in this mandatory introductory package a core curriculum requirement in social environmental stewardship? When the organization's environmental policies are being designed, the core curriculum in environmentalism should be one of the policy criteria.

Students get involved in the environmental movement for many different reasons. A common goal for such students is their wish to become involved in 'real-life' projects and research. No one wants to work on a project which has no relation to the 'real' or outside world. The most effective student projects involve partnerships between faculty, staff and students. In these collaborations, students have the possibility of 'internalizing' or owning their projects, taking the chance to examine their own organizations and the surrounding communities. The complex interrelation of environmental, social and economic concerns is perfectly demonstrated in most eco-projects. Students have more commitment and willingness to get involved than we have historically given them credit for and offer us lessons that we should investigate so that we can learn from their successes.

In the earlier chapters I briefly mentioned a student group on the Rutgers Campus that I (and my Rutgers colleague, Ray Ching) was working with to design the environmental policies that the university still uses today. Students for Environmental Awareness (SEA) and their graduate student advisor, Eric Zwerling, gave new meaning to the words student activist. Without SEA, there would not be a university-wide environmental policy nor most of the environmental programs that exist today at Rutgers. In some ways, the on-campus environmental activities of the purchasing department were released and given a platform due to the SEA efforts to institutionalize environmental policy.

I first met Eric Zwerling in 1990. Eric was a PhD student at Rutgers and directed the Noise Program, within the Environmental Science Division. I was fresh from a waste management contract writing tour. Eric and many students were lobbying every student, faculty and administrator they could find to pass a resolution to install an environmental program and policy on campus. At the same time, I was discussing similar ideas, but with a purchasing undertone to my argument. One day in 1990, Eric and his students had made the rounds through our purchasing department and they were discussing recycled content paper with our assistant director. I was called into the meeting and since that meeting we have continued to collaborate. SEA's best work can be seen at the various graduate schools around the country where the students who have left Rutgers and SEA are working to further their education at a higher level. The other evidence of their influence can be seen in the mountain of letters that they wrote during their push for environmental justice (on campus). The letters may be useful for others who wish to achieve similar goals, and they do give a small window into the passion that was felt by the students:

- Extracts from a sample SEA University Letter 1993
  Rutgers Recycles; Won't you join us?
  Reduce it, sort it, buy and use it.
  We need your help implementing two recycling policies that have just unanimously passed the University Senate. The Solid Waste Source Reduction Policy and the Recycled Products Procurement and Use Policy are our chance to fundamentally improve Rutgers' relationship with the environment, while creatively complying with state mandates.

  Rutgers is a huge operation, and tremendous benefits would accrue if we bought and used things with an eye for waste reduction, maximized our collection of recyclables, and specified recycled content products as our procurement choice. Achieving those goals alone would have significant impacts, however, we are also entrusted with training the future leaders of the twenty-first century – the students. Their future conduct will be a result of what we ask of them while they spend four years with us, and the type of example we portray, which is vital if they are to embrace these necessary practices.

  We realize that nothing can be achieved through mandates alone, and that's why we're requesting your active participation. These policies have progressed this far because of an unprecedented coalition of students, administration, faculty and staff, all working together for a common goal. Now, at the next step, we need everyone to at least understand the goals, and those who are self-motivated to step forward and join us in the education process.

- Extracts from the SEA Questionnaire To the University Community

Helping Each Other!

First We Help You

Recycled content products are available to replace many of the virgin items that you now use. The quality is high and the prices are similar, and in some cases even cheaper.

Paper waste can be reduced through actions such as double-sided photocopying, reducing two pages to fit on one page (still easily readable), short memos copied several times on one sheet then cut and distributed, and accurately targeted mailings. These actions save paper, money, waste, energy, and reduce air and water pollution.

Recycled content paper is currently available that is virtually indistinguishable from virgin paper. The RU Senate has adopted resolutions that encourage everyone to use recycled content paper whenever possible, and nowadays that's just about always. We need professors, secretaries, students, administrators, etc. to start asking for recycled every time. Recycling must be a cycle to work!

Did you know that you can buy recycled copy paper for $22.75/case from the RU contract?

Then You Help Us

The Source Reduction and Recycling Committee has been set up to consider ideas and then implement those that are reasonable. We need people to help us in our deliberations: input on old ideas, fresh new ideas, ideas about educating and involving as many people as possible (a main goal), etc.

Orientation education. We will be conducting educational sessions at the orientation of all new students: undergraduates, graduates and transfers. We will also conduct sessions for faculty, staff and administrators. In other words, everyone who makes up the Rutgers community is asked to participate, and we're giving them the tools.

Waste audit. We're going to conduct a major waste audit to look at areas such as these: major single sources of uncaptured recyclables; major inconveniences impeding compliance; areas of chronic or significant non-compliance; staff implementation of recycling policies; and suitability and placement of current receptacles.

In addition to SEA's work, many other students were involved in a variety of student projects and internships with me. Two of my favorite projects involved collaborations with two particular students: Shari Stern researched and authored *The Rutgers University Recycling and Recovery R-Plan Report*, and Marissa Perry researched and authored *Investing In Rutgers: The Resource Recovery Facility Report*, some extracts from both of which are given below:

- *The Rutgers University Recycling and Recovery R-Plan Report*
  The R-Plan is a proposal to make recycled content paper available to Rutgers University at a lower cost and improved quality while stimulating recycling efforts and promoting reduction of natural resources. Using a 'closing the loop' approach it incorporates the familiar 'three R' approach to recycling – reduce, reuse, recycle plus a fourth component – repurchase. The development of the project involves understanding the process of the university's purchasing decisions, maintenance co-ordination, the federal, state, and local laws regarding solid wastes, and the paper industry market ...
  Federal, state, municipal, and the university have all identified the importance of recycling in solid waste management as well as the need for strong recycling markets. Initiatives for procurement programs for large organizations have been set by government officials but consumer demand for the quality, dependability, and lower costs associated with virgin materials has infringed the recycled market. A program to unfold the complications with cost, quality and the un-education of purchasers is in demand.
  The purpose of the R-Plan is to tailor a recycling and repurchasing program to the needs of Rutgers University. Although this program is specifically designed according to the infrastructure of Rutgers, it may be incorporated into other large purchasing agencies. The basis of the R-Plan is to drive the recycled paper market by designating the university's paper resources to a supplier and use the resulting financial benefit to purchase recycled content paper from that same source ...
  ... According to the R-Plan process, the white paper would be separated from the mixed paper recycling stream, sold by a waste hauler to a paper mill for the market price, and integrated into a mix of white bond paper to be remanufactured into a high quality recycled paper with a determined percentage of post-consumer content. Each institution would be 'reimbursed' for repurchasing its own remanufactured paper by a percentage discount of its market value by the volume generated by the university. The intention is for the purchasing agent to be driven

towards this paper over virgin due to a combination of its competitive or lower cost, quality performance, and education pertaining to natural resource conservation.

- *Reinvesting in Rutgers: The Resource Recovery Facility Report*
Reinvesting in Rutgers is a program whereby Rutgers University would invest in the Pilot Composting Facility designed by Dr Melvin Finstein which has not yet been built due to funding problems.

  The premise of the program is that a larger-scale composting facility would be built based on the functioning of the pilot to accept all of Rutgers' organic waste thereby diverting it from other more expensive means of disposal. Therefore, investing in the Pilot Composting Facility would decrease the expenditures made on waste disposal. In addition, a facility like this could lead to universities, businesses, or organizations taking care of a large percentage of their own waste stream, recycling the balance and reducing their dependency on incineration and landfilling. In short, there may not be much left to take off campus to dispose of if a facility like this is operational.

  Activities included in the project:

  - Gather data necessary for an economic analysis of waste disposal costs at Rutgers.
  - Examine what is currently spent on the disposal of organic waste and the costs for the future.
  - Estimate the cost of building the full-scale composting facility.
  - Determine if the premise of the Reinvesting in Rutgers program is viable.
  - Incorporate design considerations into the economic analysis.

  Note: The compost facility being planned is an in-vessel Dutch tunnel system.

These extracts give a small taste of the excellent work that was produced by two very dedicated students who both graduated from Rutgers and went on to graduate school.

The best thing about being a tutor for a student internship is assigning projects which are relevant and can be implemented at your organization: as a professional you can have students working on 'real' projects where the results that they come up with can be used and implemented. It is worthwhile for many more of us to participate in the education of our young people with real-life projects with real-life results. The students who have worked with me have been assigned intense projects, but the rewards that our organization received as a result of their work cannot be measured with grades alone. We were able to prove theories through their research that members of the purchasing department did not have the time to prove with the busy schedules that they all have.

There are a whole range of other initiatives which involve students in environmental action. CEED and BEES are two examples which are outlined below.

- CEED (Community Environmental Educational Developments) is a student-run organization at the University of Sunderland in the UK. CEED is involved in adult education programmes which lead to a wide range of local people being equipped with the skills and knowledge to make a positive contribution to their personal, local and global environment, possibly leading to qualifications in the longer term. CEED also undertakes practical conservation projects with schools, and has been instrumental in creating the Student Declaration for a Sustainable Future.
- BEES (Building Environmental Education Solutions, Inc.) is based in the US. It develops environmental education programs which are multidisciplinary, inquiry-based, and customized around a local issue, and involves 'middle to high school students in an in-depth and hands-on examination of an issue, incorporate many perspectives, and include the use of technology tools, exposure to career opportunities, and practice in using workplace readiness skills' (BEES Web site).

    BEES is supported by the American Re-Insurance Corporation under their Re-wing program and is a broad-based coalition of resources from the academic, corporate, environmental, media, government, and community sectors which connects schools to local resources – experts, curriculum and resource materials, facilities, and funding sources to implement and sustain their programs.

My first contact with BEES was during a nine-week pilot project involving four schools in the greater Princeton and Trenton, New Jersey, area: Trenton Central High School, Granville Academy, Hunterdon School of Princeton and Hunterdon Central Regional High School. During this initial pilot program, the site selected for study was a contaminated and abandoned seven-acre site in a residential section of Trenton, NJ. An elementary school and a fire station were in the immediate neighborhood. In each week of the project, a session was held outside the classroom. These students were not playing games, they were doing high quality work which included:

- touring the abandoned site
- media and manufacturing processing
- site characterization and evaluation techniques
- remedial design and clean-up
- biotechnology in site remediation
- land reuse and planning.

One of the most impressive BEES projects was the Cooper River Study conducted by the students at Camden High School. Working with high school faculty and the BEES administrative team, the Cooper River Project involved dividing students into research teams to document and report on the history and ecology of the river, as well as mapping and analysis documentation. This level of academic work and commitment was impressive and my own view is that if all students could work as well as these, our future would be in good hands.

Students and the future

Our students today will be working and living in the future that we are designing for them today ... this is an overused statement because, without our meddling, the future automatically goes to the next generation. The difference we can make to our students' futures is to get them involved now while they still have an opportunity to participate in the creative planning and education phase of their lives: the 'real-life experiences' we always talk about. Once they become adults with job security, career progression, mortgages, car payments and all the pressures life brings to them, it might be too late to discuss sustainable development and environmental preservation strategies. This might be the problem we are facing with our current generation. We were not brought up with a collective desire or opportunity to get involved in the environmental movement, so most of us still believe that it will take care of itself or that the few people who do care will take care of things for us. More of us need to take a moment to investigate what our students know about our environmental future and have the courage to ask them 'What would you do and how would you do it?' The answers you get may surprise you and you will begin to learn that the gatekeepers of the future have already begun their journey.

International university programs

The university occupies an important place in bringing about environmental changes because, across the world, the university is the principal gateway through which the vast majority of our future leaders and decision-makers will pass. What is significant is the timing of that journey – they will usually pass through this gateway *before* they join a profession or organization or sign up as an employee to a company. If at this early stage these young people can acquire environmental knowledge and understanding and deep personal commitment, they can take this conviction with them into the organizations that they will later join. Still later, when they reach positions of influence and power, it should underpin all the decisions they make and the responsibilities they undertake.

University presidents and vice-chancellors in the USA, UK and Europe, along with their senior academic and management staff, are well aware of the role of their universities in this respect. A number of high level university environmental initiatives – declarations, resolutions and charters – have taken place world wide-since 1990, including:

- October 1990. The Talloires Declaration of University Leaders for a Sustainable Future (originally named University Presidents for a Sustainable Future); Tufts University USA at its international conference held in Talloires, France.
- December 1991. The Halifax Action Plan for Universities from the conference on Creating a Common Future, Nova Scotia, Canada.
- August 1993. The Swansea Declaration of the Association of Commonwealth Universities, Wales.
- Autumn 1993. The Copernicus (Co-operation Programme in Europe for Research on Nature and Industry through Co-ordinated University Studies) University Charter for Sustainable Development of the Conference of European Rectors.
- November 1993. The Kyoto Declaration of the International Association of Universities, Japan.
- July 1995. The Student Charter for a Sustainable Future drawn up by a conference of the Student Unions of the United Kingdom, Sunderland, UK.

In the UK, this movement was given added impetus by the appointment of a government committee to study environmental education in further and higher education. The committee became known as the Toyne Committee after its chair, Professor Peter Toyne, vice-chancellor of Liverpool John Moores University. It reported in 1993 in a substantial document entitled *Environmental Responsibility*, often referred to as the Toyne Report (DfE et al).

In its opening paragraphs the Toyne Report stated the reason for its existence:

Against a background of ... heightened public awareness and concern, [the Committee] had been asked to examine the present state of environmental education in FHE in England and Wales, and to make an assessment of what needs to be done, now, to provide the workforce with the knowledge, skills and awareness which it will need to assume greater environmental responsibility.

Naturally, the committee focused most of its enquiries on educational curricula and courses but it also noted the fact that these courses were taught in the context or setting of large institutions, often with multimillion pound turnovers, substantial amounts of buildings and land, consuming very considerable amounts of energy and water, and involving significant transport use. It observed that institutions should 'practice what they teach' and should therefore 'develop a comprehensive environmental policy covering all aspects of environmental performance'.

Three years later, a follow-up study, The Toyne Review, was conducted (Ali Khan 1996). This noted that not a great deal of progress had been made, and so made a more pointed recommendation, as follows:

Within three years all further and higher education institutions should either be accredited to, or committed to becoming accredited to, a nationally or internationally recognised environmental management standard, such as the eco-management and audit scheme (EMAS) ...

The implementation of environmental policy is at an early stage in both the further and higher education sectors. Most progress has been made on improving housekeeping practices, particularly in areas where there are obvious cost savings, such as energy efficiency, or where the green ticket can help institutions introduce otherwise unpopular measures, e.g. car parking charges. Less progress has been made in areas such as purchasing. (Ali Khan 1996)

There are also various other international programs in higher education which deserve further attention. We outline below just two: the University Leaders for a Sustainable Future (ULSF) program, and WWF (the World Wide Fund for Nature) UK's Higher Education program.

University Leaders For A Sustainable Future (ULSF)

Dr Tom Kelly recognized, even before I did, that purchasing departments acted as gatekeepers to all the materials, goods and resources that entered an organization. He realized immediately that the role of purchasing was universal, and that the tools of environmental contract management could be applied in any organization of any size in any sector. He saw how it related to his own program: he was the director of the Secretariat of University Presidents for a Sustainable Future, as it was then known before becoming, in 1995, University Leaders for a Sustainable Future (ULSF).

The ULSF's inaugural statement is known as the Talloires Declaration because it was made at the culmination of the international conference on The Role of Universities in Environmental Management and Sustainable Development, which was held in 1990 in Talloires, France, at the Tufts University European Center based there. The Talloires declaration is a collaborative statement by 31 university leaders and international environmental experts representing 15 nations from around the world. The declaration is a framework for action which highlights the roles and responsibilities of universities in supporting environmentally sustainable development and advancing global environmental literacy.

The declaration was the very first agreed public commitment by university presidents that sustainable development and environmental preservation should be a priority concern for the university sector. It begins as follows:

Signatories,
We, the presidents, rectors, and vice-chancellors of universities from all regions of the world are deeply concerned about the unprecedented scale and speed of environmental pollution and degradation, and the depletion of natural resources. Local, regional, and global air and water pollution; accumulation and distribution of toxic wastes; destruction and depletion of forests, soil, and water depletion of the ozone layer and emission of 'greenhouse' gases threaten the survival of humans and thousands of other living species, as well as the integrity of the earth, the security of nations, and the heritage of future generations. These environmental changes are caused by inequitable and unsustainable production and consumption patterns that aggravate poverty in many regions of the world.

We believe that urgent actions are needed to address these fundamental problems and reverse the trends. Stabilization of human population, adoption of environmentally sound industrial and agricultural technologies, reforestation, and ecological restoration are crucial elements in creating an equitable and sustainable future for all humankind in harmony with nature. Universities have a major role in the education, research, policy formation, and information exchange necessary to make these goals possible.

University leaders must provide the direction and support to mobilize internal and external resources so that their institutions respond to this urgent challenge. (ULSF 1998)

Some of the key items on the list of actions that the group pledged to undertake are shown below, abbreviated for simplicity:

1 Raise awareness at the highest levels ... of the urgent need to move toward an environmentally sustainable future.
2 Encourage all universities to engage in education, research, policy formation, and information exchange ... to move toward a sustainable future.
3 Establish programs to produce expertise ... to ensure that all university graduates are environmentally literate [so that they can be] ecologically responsible citizens.
4 Create training programs to enable university faculty [academic staff] to teach environmental literacy at all levels.
5 Set an example of environmental responsibility by establishing institutional ecology policies and practices ... demonstrating environmentally sound operations.
6 Encourage interdisciplinary research ... to promote sustainable development. Collaborate with community and non-governmental organizations to find solutions to environmental problems.
7 Collaborate for interdisciplinary approaches ... [by convening] university faculty, administrators and environmental practitioners to develop ... [environmentally sustainable] operations.
8 Enhance the capacity of primary and secondary schools [by establishing] partnerships to ... develop ... interdisciplinary teaching about population, environment and sustainable development.
9 Work with national and international organizations to promote a world-wide university effort toward a sustainable future.
10 Establish a secretariat and a steering committee to continue this momentum and to inform and support each other's efforts in carrying out this declaration.

This summary is taken from the version of the declaration which was updated in 1994, following the United Nations Conference on Environment and Sustainable Development (UNCED) in Rio in 1992, to reflect progress and redirect its focus.

From its splendid start in 1990, the Talloires Declaration has grown to over 250 signatories from 43 countries across five continents. Signatories are divided equally among low/middle income countries and high income countries and represent both large and small public and private colleges and universities, community and technical colleges, and research centers. Affiliate members of the declaration also support universities in their work and these members include government agencies, corporations, and non-governmental organizations.

> The Talloires Secretariat (operating from the Center for Respect of Life and Environment (CRLE) in Washington DC, USA) has a unique national and international mission in promoting and supporting academic leadership for the advancement of global environmental literacy.

My own work with ULSF has been to share information about environmentally responsible purchasing strategies throughout the ULSF membership, based on the ULSF outline and strategy. This strategy focused on the following four areas for action within the universities:

1 *Curriculum*. In all subject disciplines, academic staff need to review their subject content in the light of its relevance to environmental issues. Upon such consideration, all disciplines turn out to be relevant, sometimes in surprising ways.
2 *Research*. The advanced specialist knowledge of university level research contributes hugely to the search for solutions to environmental problems. By increasing the number of environmental research projects, danger to the environment will move higher up the agenda of concern.
3 *Operations*. This book arose from scrutinizing the everyday operations of one large university. Multiply that by every college and university in every nation and you are talking of a mighty force for change – and one that is equally applicable to corporations, business and industry, local government, public agencies and NGOs.
4 *Partnerships*. The research and academic partnerships that flow in and out of every department in a university into industry and business, local communities and government are superb channels of communication, exchange and influence. These links and partnerships can muster huge amounts of expertise and knowledge.

The focus of this book falls into the third category – operations – and draws on the experience of one university department to convince others to use the power of purchasing as the weapon of universal change. This message has been spread internationally by many of us in an effort to put environmentally responsible purchasing action points within the programs of the Talloires signatories. For example, the paper recycling program in Cusco, Peru, which was described earlier, was a direct result of the partnership with the ULSF. So too were a whole series of visits to the UK university sector, speaking at conferences and in specially convened seminars on how universities and colleges might improve their environmental performance through their purchasing and contract writing.

WWF UK's Higher Education program

WWF UK's work in higher education focused on publishing teaching materials and textbooks which supported the environmental curriculum at university level, all of them written by university teachers who had themselves devised the new course elements and practical exercises. These materials provide information and updating for lecturers teaching within those subject disciplines which have not sufficiently emphasized their environmental impact (such as architecture, business, economics, transport studies), or which can help us understand the environmental predicament (such as philosophy, history, literature) or illuminate concepts vividly (art, design, animated film making).

Allied to the work with teaching staff has been the work with students, harnessing their energy and dedication by pioneering student projects with a strong environmental focus. For example, WWF UK sponsored a national animation competition (run by the Royal Society of Arts and now in its ninth year) which invited student film makers to make punchy and entertaining environmental films. The winning films are now shown world-wide in cinemas and on television.

Student projects on environmental performance

Although curriculum-based work has been growing, the longest-standing and most widespread project work by students, found in most universities in the UK, has been their contribution to 'greening' the performance of their own university. The practical knowledge and experience gained from student projects to monitor and curtail their college's use of energy, water, paper, private transport, etc. is a highly transferable knowledge and skill, and one immediately applicable to their future destinations after they graduate, in the workplace and the home.

Sunderland University has created a collaborative network of other international university signatories: the Talloires Environmental Citizenship Network (TECNET). The network universities, headed by Sunderland University, have gained funding for several major international projects.

The ALFA project has taken the declaration to new heights. ALFA is the European Commission's exchange program between universities in the European Union and in Latin America: ALFA stands for Amérique Latine – Formation Académique. ALFA has two parts which combine the curriculum and operations to achieve the goals that have been a constant feature of this book – the environment, the economy and equity, as outlined below.

- The first project is funded by EC Directorate General 1 (DG1), External Economic Relations Directorate for Latin America. In the current global economy, there is a tension between the need for socio-economic development and the need for conservation of natural resources and environmental quality. This tension is particularly acute in developing countries such as those in Latin America where environmental problems may possibly overwhelm development gains. Complex issues cut across disciplinary boundaries, making it difficult to marshal the necessary academic skills to tackle them unless a multidisciplinary team is assembled.

  This particular project aimed to create interdisciplinary case studies as a vehicle to focus on key issues for the twenty-first century: economy, environment and equity (in the sense of social justice). The project is targeted at university undergraduate education on environmental issues.

- The second project which gained EC funding (this one from Directorate General XI) was for an Environmental Education and Environmental Reporting Network. This project aims to combine environmental information and the World Wide Web for educational purposes. With partners in Spain, Italy, Germany and Greece (all signatories of the Talloires Declaration), the aim is to create a multilingual Web-based environmental education resource.

  The project team has created a clearing house of environmental education sites grouped under subject headings suitable for use by teachers and pupils around the world, covering environmental reporting by businesses, social and ethical reporting, sustainability, listings of environmental newspapers and magazines and international government environment departments. It highlights subjects such as natural resources, biodiversity and forestry, earth sciences, population, transport, science and technology.

The Internet

This book offers many contact addresses on the World Wide Web (see list of contacts at the end of the book), so that you can read for yourselves the full text of the items we have quoted. You may explore further the many other aspects of any particular site, or you may follow the links to other sites, and other Web pages elsewhere. It is a unique form of searching and finding, unlike any other form of communication on earth.

The Web offers a wealth of information for those who want or need it and the creation of a Web page is a generous act in itself. Information and knowledge is displayed and freely available to all comers. The international exchange of such knowledge and expertise can be achieved at a fraction of the cost of post, telephone, fax, photocopying, publishing, conferencing and visiting. Not only is it cheaper and quicker and more far-reaching, but it is still also (so far) characterized by the spirit of generosity, co-operation, sharing and free exchange. In the same spirit that this book urges local economic development and support for the purchasing of goods and services, it urges a far more global approach to the acquisition of knowledge and expertise.

It is particularly significant that the companies, organizations, charities and NGOs, professional bodies and associations that we have listed for you are all bodies which realize that the Web is an essential form of information provision. These bodies are aware that their readers are far-flung, scattered throughout and across countries, and distributed among the most enormous range and variety of businesses and institutions. The Web enables this widely and thinly scattered population to come together in significant numbers – scattered minorities can become a mass movement in cyberspace.

Coming full circle, among the thousands, and indeed hundreds of thousands, of individuals who act as purchasing professionals there is a growing awareness of the environmental element of their role. They recognize the pivotal place that they occupy in any organization, the responsibility that they hold for environmental impact, and also the possibility that they might make a positive contribution to changing the attitudes and culture of our society. The evidence of this trend is available in documents, publications and the stated concerns of the professional associations of purchasing and procurement: many of these are also available on the Web through, for example, CAPS or CIPS sites.

Bjorn Stigson, the president of the World Business Council for Sustainable Development (WBCSD), has recently written:

> To date, the focus of most companies' actions has been on the production side of the equation. But this is changing. Increasing attention is now being paid to the consumption side – and this is certainly an area where the WBCSD is actively engaged: for example through its ongoing work on Sustainability in the Market ...
>
> I believe that the multiplier effect of companies managing their supply chains along such lines can be another significant driver to promoting and cascading the practice of eco-efficiency more widely in industry, in particular to smaller enterprises. (Russel 1998b)

Conclusion

There is still much work ahead but these are the kind of relationships that keep you in the game for the long ride. Partnerships and collaborations are one of the keys to global environmental success. You do not have to look far to find working relationships such as those described, but you have to be prepared to take a journey – either a virtual journey through the Web or by stepping away from your desk, talking to new colleagues and taking on new challenges ... you never know where your next adventure could take you.

7
Corporate Environmental Initiatives

This chapter leads us on from a study of the activities within one particular organization (Rutgers University), and the external contacts which help to sustain that work, out into the wider world, giving us a chance to look at what the corporate giants are doing about their environmental performance. We look at what we can learn from these companies, at what – perhaps – they might learn from us and, more importantly, how we can work with these companies for mutual benefit and to contribute to global sustainability.

It is increasingly understood that, to be successful, companies will need to do more than just respond to new environmental legislation, and will have to start being much more proactive in this field. For example:

> The move toward sustainable development has been building for twenty years, and as we approach the millennium, the time has come for corporations to begin preparing for a sustainable future. They must become proactive and help to lead the way, or risk being swallowed in a reactive, heavily legislated, and adverse business environment.
>
> The successful company during this transition will need to be nimble. Communication will be critical as the waves of information, change, and opportunity build. Ferreting out trends and developing issues will also require external business networks such as The Global Business Network and the Future 500. Preparing for the future will require forward thinking.
>
> By taking the lead, corporations can move society toward sustainability far more efficiently and with less turmoil than governments and legislation. There will be many challenges, and perhaps many more opportunities. Companies with the vision to help shape the future will

have a much greater chance of recognizing opportunities than those who merely react to the changes. (Forrest 1996)

Sustainability: is this word overused? is it understood?

Not everyone has the same understanding of what sustainability means. The most common definition of sustainable development is as follows:

> Humanity has the ability to make development sustainable – to ensure that it meets the needs of the present without compromising the ability of future generations to meet their own needs. (WCED 1987)

When applied to economic strategy, it has been described as follows:

> Sustainability is an economic state where the demands placed upon the environment by people and commerce can be met without reducing the capacity of the environment to provide for future generations. It can also be expressed in the simple terms of an economic golden rule for the restorative economy: leave the world better than you found it, take no more than you need, try not to harm life or the environment, make amends if you do. (Hawken 1993)

For the purpose of this book, we shall accept these definitions. On this basis, the implications of sustainability for the corporate sector are profound:

> As we look deeply into the question of how to make a better world, every direction leads to one essential fact: most of our current activities are not sustainable. In spite of the laudable efforts for conservation, recycling, and forms of socially responsible business, the end result falls far short of sustainability. In our current market dynamics and resource utilization, we are, by definition, seeing to our own demise.
>
> Our global transition to sustainability will have to be fundamental and profound, and undoubtedly a little painful. For most citizens of the world – and particularly for those in first world countries – it will mean tangible changes for every single one of us. The sooner we act, however, the less painful it will be. IF we act now. (Nelder 1996)

This chapter will not attempt to describe or to explain what is going wrong in the world of non-environmentally friendly industry practices and the effect these companies are having on our dwindling resources, greenhouse gases and other unnatural disasters. These are real concerns and many journal articles and research documents have been written to prove that these occurrences are indeed happening the world over. There is a list of books and journal articles listed at the end of this book, should you wish to research and pursue environmental issues further.

Business leaders are clearly beginning to recognize these issues. Derek Wanless, Group Chief Executive of the NatWest Group, Chairman of the UK Government Advisory Committee on Business and the Environment (ACBE), member of the World Business Council on Sustainable Development (WBCSD), has said:

> Firstly, [there is] the integration of environmental issues into the core activities of business. Such integration and public reporting on progress are, I believe, necessary first steps in achieving the environmental improvements which we all seek.
>
> And secondly, [there is] the case for long-term partnerships across all sectors. If we are to achieve positive improvements in environmental management, I believe that action must involve business, regulators and government. Ultimately, as representatives from local government, business, the voluntary sector and academia, our collective aim must be to help each other move towards the goal of sustainable development ...
>
> In conclusion, the challenge presented to us by the environment has clearly grown from initial issues surrounding local pollution to concerns about global threats and choices. The sustainable development challenge presented to businesses has grown in very much the same way. Businesses started with simple, technical environmental issues and initiatives and now look at corporation-wide threats and choices. (Wanless 1995)

One example of how seriously businesses are beginning to take the sustainability message is the agreement between five of the world's major banks and the UK in a document called *Banking and the Environment. Statement by International Banks on Environment and Sustainable Development*. This includes the following statement:

> ... The United Nations Environment Program Advisory Committee on Banking and the Environment [aims to produce] a policy on the role of banks in terms of improved environmental performance leading ultimately to sustainable development.
>
> The United Nations Environment Program acknowledges the potential influence of the world's banks in changing business practices. As financiers, bankers have a unique role to play ... International co-operation

and long-term planning are key to the objective of balancing economic and environmental considerations.

This agreement among international bankers is a valuable tool for anyone developing environmental policies and programs. It is very rare to see co-operation from a powerful group in order to agree on anything and, for whatever it is worth, we should take this document and see how much commitment we can get out of the signatories, the individuals who designed it and released it. In most of the recent bank mergers in the United States, many communities and community-based action groups have rallied together and got the banks to commit to investing in the community as part of the agreement allowing several different banks to merge. This recently happened with the First Union/Cores States Bank merger which will benefit communities in New Jersey and Pennsylvania. The spirit and determination shown by these communities could give them the drive to include LA21 commitments as well. In the increasing global connection of our current world, international banking, local authorities and community groups are all potential partners.

Relating to corporations

As purchasing professionals, it is our job to know where the resources that it takes to manufacture the products and services we procure are to be found. If we do not concern ourselves with the danger of depleting the natural resources that are being used, it means that we are tacitly accepting such practice and allowing our contract vendors to do the same. It is not only actions that are culpable: failure to act can also be a crime. We have to try to discover how involved and knowledgeable our vendors and companies are about their own corporate environmental responsibilities. If we procure materials and services from the vast multinational and international corporations, and do not express our concerns to this corporate world, it allows them to discount damage done to the environment because it doesn't show up on their 'bottom line'. This discounting or ignoring of environmental damage is made possible through 'externalizing' the loss. For example, if land or rivers are polluted, if soil is exhausted by intensive cropping, and if the organization responsible for the damage bears no costs of cleaning up or of repairing the destruction, it has managed to keep those costs 'external' to its balance sheet. Those costs will fall elsewhere. It may be left to governments (national or local), and finally to individuals and communities, to foot the bill for the cleaning up or the repair. Equally, it may be local communities who suffer damage to their health or their livelihoods as a result of corporate activity that destroys or harms the environment which provides their livelihood and their home.

What is our relationship with the corporate giants? Let's face it, we need industry and industry needs its consumers. No one wants companies to go out of business due to environmental regulations. Smart business in most cases faces up to questions of environmental and social ethics. In many cases, our business leaders are extremely concerned about the same environmental issues that our communities are concerned about. The problem is that until relatively recently there has been little publicity about their environmental initiatives and how these affect the community in positive ways.

Articulating the facts, reporting them widely and making sure the consumer is part of the ultimate success of their environmental agenda is one of the keys to creating an informed public and improving a company's public relations. Environmental reporting has traditionally been a voluntary method of communicating environmental performance to an organization's stakeholders. However, more recently there has been a debate over whether environmental reporting should be made mandatory. Denmark, New Zealand and the Netherlands have already started introducing legislation on environmental reporting. The voluntary European Eco-Management and Audit Scheme (EMAS) also requires that environmental statements be produced, although the International Standard ISO 14001 does not specify that they be made publicly available.

The large companies are particularly active in the field of environmental reporting, preparing substantial and colorful Company Environmental Reports (CERs) and publishing summaries on the Web for anyone to inspect. There is a Web site which lists company environmental reports, both in printed form and on the Web. The reports are listed alphabetically, by sector and by country and cover legislation on environmental reporting, reporting guidelines, details of environmental accounting papers, environment and financial performance, environmental management systems (such as ISO 14001 and EMAS) and journals dealing with business and environmental issues, as well as links across to a social and ethical reporting clearing house. The Web site can be found at <http://cei.sund.ac.uk/envrep/reports.htm>

When we began the process of the greening of Rutgers University, one thing that we had to overcome was the negative historical statements made by industry about environmental preservation movements. This was the clear message that we were getting from the media and from several congressional leaders. We heard that 'environmentalism is bad for business', 'environmental regulation will cost industry millions of dollars to make the transition', 'people will lose their jobs, and companies will go out of business'. It is also very clear that industry has a powerful influence over our political system: so great is its power that it is amazing that any environmental laws are passed at all. The political process includes the representation of key industries, and in this way industry has made sure that it will have a say in what laws are eventually passed. In addition, industry is rich and powerful enough to spend millions of dollars lobbying politicians to vote against environmental regulations that they feel will hurt them financially. As industry is armed with millions (sometimes billions) of dollars worth of research, it is a formidable task for individuals or groups to prove that industry is wrong and to challenge and confront its messages. This political lobbying effort by industry is rarely heard and almost never seen by ordinary members of the public. Media coverage does not seem able to get the story across to the public in a way that makes them feel outraged at what is going on behind the scenes and galvanizes them into taking some action.

The work ahead

Identifying a strategy to green industry and then enforcing it will be a monumental challenge to our civilization. Industry has enormous power over the value systems of entire populations. If industry wants to shift our thinking towards buying bigger and faster without consideration for environmental consequences, it will concentrate its efforts, research and financing in this direction. As multinational corporations have the ability to deliver massive global marketing strategies, the average citizen appears to be locked into the psychological grip of industry. Breaking through this stranglehold to demand that companies should become environmental and sustainable is an enormous task, but some companies are doing some of the things that can be done. As purchasers, as consumers, as stakeholders, we actually have important and substantial power over industry if we decide that we wish to exert it. We can then be a significant force in pushing for these greening transformations.

Examples from large corporations

The successful merging of economics, social responsibility and environmental preservation is rare but, if done successfully, it can act as an important role model for any business world-wide. Since this combination is a rarity, the light only shines on a few who have made this transformation. The companies that have made this combination a priority have been able to reap the financial and social benefits that come with it. Sustaining and supporting the spread of this type of system will take the continued co-operation of industry leaders and an environmentally literate body of consumers. Green industry makes good business sense and it gives our natural resources a chance to thrive so that future generations can enjoy and use them.

As far as we know, most of our industry leaders are good, decent, caring individuals who live in our communities. We have no reason to suppose that evil purposes and a lack of interest toward the environment rule in the boardrooms of industry and business. At worst, they may check in their lofty ideals along with their coats when they meet for business. But now a new round of interest in environmentally sensitive business needs to be sparked by the consumers of the world. Purchasers for institutions are, and will continue to be, among the largest, and therefore most influential, consumers for these industries. Greening industry is a two-way street. Industry will make environmental preservation and the protection of resources into company policy if we, as consumers, as shareholders, as stakeholders, demand it. Several industry leaders have listened to this call and a few have made the transformation to 'green business' with surprising results. In my role as a purchasing official, I have seen first-hand examples of corporate environmental leadership. The following examples prove that industry can make a difference to its customers (both directly and indirectly), to its supply chain and to the local economy. However, what is *said* and what is *done* are two completely different things and so we must seek out examples of companies who 'practice what they preach'. The examples given below of B & Q, Somat, Steelcase Inc., GreenDisk Inc. and Hewlett Packard show what we can begin to expect from large corporations.

B & Q, UK

B & Q is the UK's largest home and industrial DIY supply chain and it has a history of creating and implementing creative, thoughtful, environmentally and socially responsible strategies as part of their purchasing and sales mission. Two things brought B & Q to my attention:

- Their popular book *How Green Is My Front Door?*, published in 1995. This was B & Q's second environmental report.
- An inspiring lecture given by Dr Alan Knight, B & Q's Quality and Environmental Controller at the Conference On University Purchasing (COUP) held at Anglia Polytechnic University, Chelmsford, UK, in September 1997.

How Green Is My Front Door? is a well-documented history of B & Q's involvement in environmentally responsible purchasing and quality control. Alan

Knight, with a hands-on account of what B & Q is doing around the world to institutionalize responsible practices, brings the book to life with dozens of real-life examples and case studies. In short, Alan Knight's environmental team has complete knowledge of all the purchasing resources that B & Q has around the world. They have systematically traveled to these locations to invoke socially and environmentally responsible changes. For example, B & Q has funded a community-based forestry project in Papua New Guinea and is actively working with other suppliers in developing countries to improve the working conditions particularly in factories in India and in the Philippines. As I have mentioned in previous chapters, the rush to globalization has a real impact on the lives of local people, and the resources of other nations are paying the price. Alan Knight has taken B & Q purchasing professionals to these international locations and shown them what their purchasing processes have done to these communities. The purpose of this eye-opening experience was the next step in B & Q's strategy to involve their scientists, local community social work officials and purchasing officials in the transformation of current practice into responsible practice.

In addition to these achievements, B & Q has a wealth of information available which shows their corporate environmental actions. Their environmental philosophy states:

> Our most significant impact on the environment is caused by the 40,000 products we sell in our 283 stores. We are committed to reducing that impact.
>
> B & Q decided in 1990 that the company needed to know what issues should be addressed (by research), decide how to address them (by devising a policy) and then formulate an action program, often with targets, and monitor results.
>
> B & Q will continue to concentrate on working with its suppliers in the UK and around the world. Suppliers are expected to continually improve their environmental performance and B & Q's grading system will reflect higher and higher expectations as years go by ...
>
> In particular, more time of the environment team needs to be devoted to the issues raised by the international supply chain. Suppliers in these countries need to be more closely scrutinized and more action plans put in place before we can even begin to feel we have fully addressed the issue of polluting factories and poor working conditions.
>
> On the product side we have a tremendous amount of work to do to ensure that all our timber products are independently certified by the end of 1999. There will be considerable negotiation with governments,

NGOs (Non-Governmental Organizations) and the timber trade to meet our deadline.

The reason B & Q started its environmental program was because the company believed that its customers and staff inherently care. However, B & Q is always first and foremost a business with a duty to its shareholders and believes it has proved that profit and the environment can work together.

B & Q suggest that their environmental achievements to date include:

- Beginning to use more efficient forms of efficient waste management and reusing or recycling in-store waste including:
  - cardboard (up to 50 percent of the average store's waste), polythene and shrink wrap, office paper, drinks cans and plastic cups, timber pallets, ink cartridges and plant trays
  - an estimated £400,000 was saved through recycling cardboard and diverting it away from general waste skips (therefore reducing the number of skip collections); revenue can also be made by selling cardboard in bulked volume
  - through energy audits in its stores, savings were made through simple actions such as setting time clocks correctly for illuminated garden centre signage and car park lights, switching off unnecessary lighting generally, reducing heat loss in winter by keeping external doors and warehouse curtains closed.

B & Q works with their supplier chain in a number of ways. Their documents state that:

B & Q, has the power to encourage its 600 suppliers to be committed to improving their environmental performance.

B & Q's policy for all suppliers was to have a written environmental policy backed up by an environmental action plan. All suppliers were instructed to complete B & Q's Supplier Environmental Audit which gave B & Q a clearer picture of relevant real and potential environmental hazards. This was a complex exercise but the result showed that the main problem was suppliers not understanding what their environmental impacts were, never mind how to deal with them.

Supplier visits, seminars, reports, thousands of telephone conversations all proved that B & Q was serious about suppliers working with them to achieve improvements in overall environmental performance. B & Q was also willing to provide the support and advice that some suppliers needed.

A grading system based on awareness of issues and a commitment to improvement rather than set standards required suppliers to achieve a certain grade or face de-listing. [Ten companies were de-listed at the end of 1994 as a result of not complying with B & Q's targets.]

B & Q merged the quality and environment departments so that both product quality and environmental issues became equally important. In the same way that customers expect a product to work and not break, they would not expect it to cause deforestation or the ravaging of other natural habitats, or even for it to come from a factory in a developing country where working conditions are appalling or where child labor is abused.

B & Q works closely with suppliers in developing countries as their documents show:

When Dr Knight, B & Q's environmental specialist, visited countries in the Far East he found polluting factories but also working conditions that were extremely hazardous. This raised the issue – should we be more concerned about destruction of forests or people dying through industrial accidents or work-related illnesses?

Issues such as poor working conditions and child labour can challenge B & Q's ethical position but they also pose commercials risks. A substandard working environment is likely to lead to poor quality products, and inadequate fire precautions increase the risk of supply breaking down. Child labour is an emotive issue which has been the subject of many column inches and most of it is coverage which no business wants to receive.

Instead of walking away from suppliers (which would have serious adverse effects on the employees), B & Q believes in working with suppliers who are committed to improving conditions. However, where a manufacturer has no interest in improving its working conditions and B & Q's business has not been crucial to the factory, then B & Q will find an alternative supplier.

Recent case studies of B & Q working with suppliers in developing countries have included:

1 B & Q sells products with the Rugmark, an independent certification scheme which ensures that hand-knotted rugs have not been produced by exploited labour.
2 B & Q is also setting up its own local independent social and environmental auditing scheme of all its suppliers in India.

3   B & Q is working with WWF and other social development groups
    in the Philippines on a project which aims to improve the manage-
    ment of capiz (a shell used in lampshades) and the health and safety
    standards of the divers who collect them and working conditions in
    the factory units.

These passages are selected from B & Q's extensive environmental action
portfolio, and there are a lot of important lessons to be learnt from reading
the company's own reports about progress in environmental and sustainability programs.

Somat

Somat has supplied Rutgers University with food-pulping equipment for
some years and further investigation of this firm revealed a much deeper
commitment to environmental preservation than could be seen at surface
level. Somat has taken the complex issue of food disposal and turned it
into a unique and successful food recycling process. Somat documents
outline their environmental philosophy, as follows:

> Caring for the environment today is quickly becoming one of the food
> service industry's most pressing challenges. With overloaded landfills
> across the nation closing every day, labor and hauling costs skyrocket-
> ing, local and state governments drafting more and tougher legislation,
> and a greater emphasis than ever before on sanitation, the need for a
> better waste reduction system has never been as great. What's needed
> are new answers, new technologies, and a better system for reducing
> the amount of waste we produce. What's needed is Somat.

> Our Roots Are In the Earth
> It was back in the late 1940s that German immigrant and inventor Kurt
> Wandel developed a small pulping unit to be used by home gardeners
> to convert organic wastes – food, leaves, grass clippings and paper – into
> mulch, or fertilizer. Wandel proved to be far ahead of his time, envi-
> sioning future applications for his technology in municipal, commercial
> and industrial applications as well. Today, we remain where we've always
> been: at the forefront of the ongoing search for solutions to the world's
> biggest problem.

> Somat equipment benefits:
> • Up to 80 percent volume reduction
> • Reduced labor cost

- Better sanitation
- Water conservation
- Improved operator morale
- Reduced silverware loss
- Processes a wide variety of waste.

These statements are part of the operational documentation that Somat presents to its customers. One of the statistics that impressed me most about this equipment is the solid waste reduction aspect ... the ability to reduce waste by as much as 80 percent. In short, the Somat equipment grinds and pulps food waste, using water in the process, along with napkins and food paper products, paper plates and cups. It then extracts the water liquid (most of the water waste is reused – recycled back into the system) and delivers a ground, oatmeal-like food waste material which is ready for composting, which is what Rutgers and other institutions use it for.

Somat's own description of how their system works is as follows:

1 *Feed Trays*. Water flow from the feed tray directs the solid waste to the pulping tank. In some facilities, a self-bussing system eliminates kitchen scraping and waste handling. Preparation and plate waste and pot wash are fed directly into pulper.
2 *Pulper*. Inside the space-saving pulper, whirling cutters turn the solid waste into a slurry. These are precision-engineered, heavy-duty cutting mechanism consisting of rotatable and removable cutting components.
3 *Silverware Recovery*. A built-in silverware trap and Somat's own trough design ensure that your money won't go down the drain.
4 *System Flexibility*. Somat's extraordinary flexibility allows side-by-side close-coupling of the pulper and Hydra-Extractor, or separation by a distance as great as a city block or more.
5 *Hydra-Extractor*. The Hydra-Extractor removes excess water from the piped slurry and produces an odor-free, semi-dry pulp. Once waste is pulped, it compacts naturally by eliminating air space and won't attract flies or vermin. The excess water that is recovered is automatically returned to the pulper for reuse.
6 *Recycled Water*. After water has been removed by the Hydra-Extractor, it is returned to the pulper, completing the closed-loop cycle.

Although this approach would only be a last resort for disposing of food materials, as the waste of food should always be kept to a minimum before it is allowed to join a waste stream, and even allowing for the marketing language, there are clearly some environmental advantages to this approach to dealing with food waste.

Steelcase Inc.

Steelcase Inc., based in Grand Rapids, Michigan, USA, has been a world-wide furniture supplier for many years. Many educational institutions in the United States use Steelcase for everything from basic to sophisticated office, classroom and conference furniture and accessories. Some examples of how Steelcase expresses its commitment to environmental preservation in their publications follow:

- *Finishing.* Installed powder coat finishing systems in Systems I manufacturing plant and File Plant in Grand Rapids, MI, and converted to water-based adhesives in Systems I manufacturing plant in Grand Rapids, MI, for production of work surfaces (1995).
- *Training.* Trained managers and supervisors to raise their awareness of environmental issues, communicate their responsibility for maintaining the highest level of environmental performance, and inform them of resources available to support their environmental efforts. In turn, managers and supervisors cascaded the information to all employees (1993).
- *Conferences.* Co-sponsored three American Institute of Architects 'Building Connections' video conferences. Targeted to architecture and design professionals nationwide, the video conferences explored state-of-the-art information on environmentally responsible design, construction and building operations.
- *Ash.* Began shipping ash, a by-product of its Grand Rapids power plant operations, to a Midwestern cement mill. The ash is used as cement filler material, reducing use of raw materials in cement production. In addition, this recycling eliminates waste sent to the landfill from the Steelcase power plant, and reduces total power plant waste by 79 percent.
- *Packaging.* Initiated a pilot project in its Grand Rapids Systems I plant to reuse corrugated packaging. Dividers made from used corrugated cost one-third of the price of new ones.
- *Transport.* Steelcase Tustin, California, facility employees awarded the State of California Governor's Transportation Award and an Orange County Transportation Partnership Award for their van pool program (1990). More than 30 employees travel up to 100 miles round trip each day in seven vans powered by propane, which burns cleaner than gasoline.

- *Recycling.* Installed distillation unit in Grand Rapids to recycle and reuse solvents used to clean painting equipment (1981).
- *Environmental management.* Hired first full-time environmental engineer (1969).

Some of the ways in which Steelcase describe how their environmental principles are translated into action are given below:

- *It Comes from the Top.* Environmental improvement begins and ends with every Steelcase employee. The drive for environmental progress is piloted by our board of directors, carried throughout the company by a cross-functional management team, and monitored daily by the Corporate Environmental Quality department.
- *A Global Concern.* Our concern for the environment reaches across the borders of the countries where we do business. (That's every country in the world.) Everyone at Steelcase understands that our environmental decisions affect the health and well-being of the planet and everyone and everything that needs it for survival. Steelcase North America, Steelcase International, Steelcase Strafor, and Steelcase Design Partnership companies strive to be good environmental neighbors. We build our manufacturing facilities to minimize their environmental impact.
- *The Materials We Use.* The materials we use to make furniture include steel, plastics, foams, wood, fabric, paint, finishes, and adhesives. For new and most existing products, Steelcase uses 30 percent recycled content steel, foams and plastics free of ozone depleting substances (ODSs), wood purchased from environmentally responsible harvesters, ODS-free fabrics, powder coatings free of volatile organic compounds (VOCs), low and ultra-low solvent paints and finishes, and ODS-free adhesives.
- *The Parts We Buy.* We insist that our suppliers provide evidence of sound environmental stewardship and compliance with all environmental regulations. This disclosure is a condition of all Steelcase purchasing agreements.
- *Preserving the Forest.* As a major manufacturer of high-quality wood furniture, we recognize that our products are created from a valuable natural resource. In keeping with our concern for preserving the earth's forests and our belief in responsible economic development, we are committed to using wood obtained from domestic sources and sustained-yield harvesters.

Steelcase's Web site contains numerous other initiatives: <http://www.steelcase.com/>

In addition to their environmental initiatives, Steelcase has developed a very basic waste reduction strategy that Rutgers University has taken advantage of: the use of packing blankets to transport desks, chairs and other furniture safely. This simple and clever strategy has resulted in the elimination of cardboard and other packaging materials that we would otherwise have to deal with as a waste issue.

GreenDisk, Inc.

GreenDisk has been my constant companion as I compute my way through life. Although not the most technical person in the world, when saving documents on a computer disk, I must feel sure that the information will be there the next time it needs to be accessed. This is all that is needed from a computer disk. However, if this basic saving of documents can be achieved in a 'green' way, that is an added bonus. The GreenDisk story is an interesting one and is worth highlighting.

GreenDisk was founded in Redmond, Washington, on April 22, 1993 (Earth Day) by David Beschen. After spending a sizeable amount of his career in marketing high tech products, David learned of the enormous disposal problem that the software industry was creating and that little was being done to solve it.

From the beginning, the focus of the company has been on providing the software industry with a highly secure, environmentally responsible method of disposing of obsolete and returned software. The result of this focus is one of the highest quality diskette products on the market – GreenDisk recycled diskettes.

The company is now headquartered in Preston, Washington, and also maintains recycling facilities in San Jose, California, with plans to open more than seven additional sites around the country and internationally.

GreenDisk's comprehensive recycling services and recycled products bring to the market a thoughtful, cost-saving and environmentally responsible solution to a growing problem...the disposal of obsolete computer hardware and software. For more details, see their Web site at: <http://www.greendisk.com/>

Hewlett Packard

Rutgers purchasing department is on the receiving end of all the requests for HP printer purchases that the campus sends in – and they send plenty. Everyone with a computer wants their own personal printer alongside it. And with the next wave of competition being copiers that print, send faxes and are networked with your computer, the office printer has a tough fight ahead.

Although the new generation of photocopiers are poised to challenge the belief that 'every computer needs a printer', Hewlett Packard (HP) has had a claim that makes them stand apart from their copier competition – and that is their recycling program. When you purchase your printer, whether it is HP or another brand, you should consider the resulting waste issues. In addition, you should consider if the company supplying you can support their equipment with well thought out environmental initiatives which address the waste issues from the outset. HP has researched these concerns and offers its purchasers an environmental package which can add value to your purchase.

Hewlett Packard describes the efforts the company makes to keep manufacturing environmentally safe:

Meeting environmental management guidelines on manufacturing is expected. We exceed them. Here's how:

- Manufacturing avoids using hazardous substances. Local, state and country environmental specifications are often exceeded by voluntary elimination of dangerous material.
- Manufacturing sites are ISO 14001 certified. The manufacturing of our HP toner cartridges meets international standards for environmental guidelines. These practices are determined by an independent auditor.
- HP stipulates strict environmental guidelines which our vendors have to meet. Vendors are also requested to audit their subcontractors to ensure they comply as well.
- Manufacturing waste is recycled. In the overall production of our toner cartridges, HP extends beyond regulations to proactively put environmentally responsible methods in place. In addition to reducing waste, waste is recycled as an energy source for other projects.

HP has made efforts to integrate environmental consciousness into every level of its LaserJet toner cartridges, while maintaining product standards. It is an awareness that extends to product design, manufacturing, distribution, operation and recycling programs. For example, they have designed and produced a printer made from up to 25 percent recycled ABS plastic in its outer casing. Their marketing material outlines the environmental benefits of this printer which include potentially diverting 6 million pounds of plastic from landfill, being lighter to transport and therefore requiring less fuel for transportation, being easier to dismantle and therefore to recycle, and energy efficiency features. More details on this and other HP products are available on their Web site at <http://www.hp.com>

The HP Planet Partners Program is another interesting initiative. This is a comprehensive approach to environmental issues affecting the daily operation of the organization and its customers. HP describes the program as follows (more details on their Web site <http://www.ljsupplies.com/planetpartners/recycling.html>)

> The HP Planet Partners Program makes recycling a lot easier. On you. And on the environment. Our efforts include:
>
> - Up to 95 percent of each returned HP LaserJet toner cartridge is recycled. Including the box and packaging material. The products made from recycled materials consist of clothing buttons, eyeglass cases and a host of other useful products.
> - Recycling programs are available in 80 percent of our world market. Markets include North America, South America, Europe and Asia. We are actively working on having recycling programs in 100 percent of our world market.
> - Since 1990 HP has diverted 18 million pounds of toner cartridges from landfill. This involves the successful recycling of over 12 million toner cartridges.
>
> HP's Planet Partners Program goes out of its way to make the recycling of HP LaserJet toner cartridges free and as easy as possible. There is a pre-paid UPS label in every HP LaserJet toner cartridge box.

One thing to look for when auditing a company for environmental responsibility is their global environmental commitment. You can be a environmental saint in the United States, but if your company has different standards in other countries ... the green label is lost. The global green mission is a vital one and should be considered part of your corporate environmental leadership evaluation and strategy. The HP program is one example of a company aiming for a green global plan in practice.

Ways forward for purchasers

There are a number of ways in which purchasers can go beyond the marketing information provided by companies keen to publicise their green credentials, including:

- indicators of environmental performance
- co-operative purchasing

• environmental advice for business.

All these approaches are described in more detail below.

Indicators of environmental performance

The idea of a common standard or benchmark against which companies can measure their improvements in environmental performance was raised in an earlier chapter, and it has already been developed in the context of environmental reporting.

An ongoing research project conducted by the environmental consultancy, SustainAbility, supported by the United Nations Environment Programme (UNEP) has resulted in the report entitled *Engaging Stakeholders*. The UNEP/ SustainAbility research program has developed a system of 'scoring' company performance on the basis of what they declare in their published Company Environmental Report (CER). Initially many environmental reports provided an opportunity for companies to blow their own trumpets, selecting their own yardsticks against which to measure themselves. Now, with so many companies jumping on this particular bandwagon, a set of common standards and measures are emerging, and the need to design the annual CER so that it scores highly on the benchmark categories is becoming the driving force for environmental change within the companies themselves. If they wish to make public their environmental achievements, they will need to put the machinery in place to make sure that they improve their environmental performance. This will be a crucial element of new purchasing policies and we will watch and support developments in this area with great interest.

Co-operative purchasing

Rutgers University and many other educational institutions have been members of the Educational and Institutional Co-operative Contract (E and I Co-operative) for many years. The key word in this contract is 'co-operative', and firms which support these contractual commitments and which have a strong environmental program (such as Steelcase, described earlier) can offer the same environmental ingredients to a large consortium of customers. If you or your organization belongs to a co-operative purchasing system, you should investigate different companies' environmental commitments and share this information with all your co-operative partners.

Environmental advice for business

There is a wide range of advice for businesses which want to improve their environmental performance, and two programmes are described below which are linked with major institutions in the UK and the US:

- MIT's Technology, Business and Environment Program, and
- WWF UK's work with small- and medium-sized enterprises.

The MIT Technology, Business and Environment Program

The need for high quality research and education in implementing projects cannot be overemphasized. The Massachusetts Institute of Technology (MIT) Technology, Business and Environment Program (TBE) was founded to help companies meet the dual challenges of achieving environmental excellence and business success. The program's mission is to elucidate a new preventive environmental management paradigm, centering on business practices and linking technological change with sound environmental management. TBE offers research opportunities and graduate-level courses within the Technology and Policy Program at MIT.

'*How* do you do it?' is always the question that is asked when you really want to implement a program and you are not quite sure what resources or plan of action you need to take. The MIT program aims to answer this need and describes its approach as follows:

The MIT Program
- Seeks to help companies shift from merely reacting to environmental laws and regulations to creating a preventive framework built on a long-term perspective.
- Fosters leadership and learning of a new pollution prevention paradigm among students and professionals.
- Strives to define and promote concepts of industrial ecology, sustainability and resource conservation.

The MIT research program includes the activities which they describe as follows:

- Analysis of corporate behavior through preparation of case studies of firms in the chemicals, consumer products, electronic and automobile industries.
- Studies identify internal and external factors which are key to environmental strategic management.

- Development and analysis of policy alternatives focusing on techno-logical innovation, products, loop-closing practices such as recycling, and industrial symbiosis.

Bringing people together to discuss and enhance their programs will always lead to further co-operation and collaborations and MIT's program pro-motes these types of partnerships in the following ways (more details on their Web site):

- Organization of conferences on research areas such as life-cycle assess-ment and product recycling.
- Conferences bring together senior people from industry, government, environmental advocacy groups, and academia to inform and co-ordinate action.
- Fostering of co-operative networks through periodic meetings of MIT Symposia on Business and Environment.
- A consortium of representatives from firms and academia attends.

The WWF UK Better Business Pack

Lack of information has always been a problem for companies wanting to be more environmentally sound, particularly if those companies are not large corporations. For that reason, WWF UK has worked with NatWest Bank to develop a Better Business Pack for small- and medium-sized enter-prises (known as SMEs). The concept behind this joint project was to increase business profits by reducing environmental impact. This reflects WWF UK's concerns not only with the natural world and its pandas, tigers and bears, but also with the man-made world of business and industry, large and small. WWF UK describes this joint initiative as follows:

> The Better Business Pack: WWF UK and the NatWest Group
> World Wide Fund For Nature (WWF) is the world's largest independent conservation organization. Its remit is to develop and implement strate-gies designed to protect species and habitats, and to reduce human impact on the environment. Landfill, pollution and the use of non-renewable energy sources, for example, are key issues affecting the environment – and key issues for business.
>   NatWest Group is a major UK financial institution and has the largest number of accounts from small- and medium-sized businesses in the UK. It operates an advanced environmental responsibility program which addresses environmental challenges and needs within the orga-nization, as well as with numerous influential outside bodies.

This Joint Initiative is carried out by the World Wide Fund For Nature (WWF UK) and the NatWest Group in response to Agenda 21 – the plan for 'sustainability' in the twenty-first century, signed by the UK Government at the 1992 Rio Earth Summit and specifically, Local Agenda 21 – the local sustainability plans being developed in the UK by many local authorities.

The Joint Initiative is a not-for-profit venture. Only the businesses that buy and use the pack will profit from it ... and, of course, the planet.

The Better Business Pack helps businesses reduce their impact on the environment in relation to waste, energy and water, transport, purchasing and supply chain. 'Sustainability' in the use of the world's resources is crucial for the future of business. Areas of concern for business are particularly:

- those which cannot be sustained indefinitely at present levels.
- those which, if sustained, will result in damaging effects on life and the quality of life.

The whole pack is firmly based on the experience of a wide range of small- and medium-sized businesses – defined as those which employ up to 250 people – in service industries, retail, distribution and production. The businesses were drawn from all sectors of the economy and included non-profit organizations such as schools, as well as commercial companies, many of them owner-managed.

Everything in the pack is designed to build on experience to bring benefit to businesses, aiming for a quick route to increased profits, with savings going straight to the bottom line.

Conclusion

To sum up, what are the things that you need to look for in an environmentally responsible industry partner? The first thing is their commitment to work within the established environmental policies and guidelines of your own organization; the second one is their concern for your need to use existing funds wisely and to obtain the same level of quality; and the third one is to look for any world-wide external commitments they might have to participate in the protection of natural resources.

In short, what are your contracted vendors doing to comply with your environmental policies and agenda, and what are they doing to make this program standard industry practice world-wide? One way of judging your potential contract partners is by studying what other companies are doing world-wide. Every environmental report you read will suggest a hundred environmental elements for you to consider using.

We have highlighted only a few of the many, many companies that are doing interesting things to reduce their environmental impact and we fully realize that there are plenty of other companies out there to tell us more. You should be on the lookout for useful information and ideas and start to create your own resource guidelines. Then you will be in a position to challenge more industries to join in and become corporate environmental leaders. If you are really successful, you may even start some new plans with your newly found environmental partners and encourage them to participate in local economic development initiatives as well.

Corporate environmental leadership is spearheaded by the hardcore dedicated business executive ... they are out there, you just have to rope them in and encourage them to become part of the big plan. In most cases, you are paying their salaries through your purchases, and the money you spend should reflect what is in your best interests and in the best interests of the environment.

8
Reaching Out to the Local Community

This chapter will add an important new dimension to everything that has been discussed in this book so far. We shall consider whether we can achieve through purchasing and contract management not only all our environmental objectives but also to go beyond these to achieve still more. Can we also develop our immediate local community by asking the local economy to supply our needs and use its products and services (which brings its own environmental benefits by reducing transport)? And what can we learn from innovative community-led initiatives on sustainability? If we are successful in creating a strong local setting for our organization, then our combined 'environment + development' package is more likely to endure and to become self-sustaining.

Sustainable economic development adds a new dimension to environmentally responsible purchasing, but it is not always clearly understood. The following definition may be useful (it is taken from the Sustainable Evansville Project, see below):

> Environmentally sustainable economic development is a process of collaborative stewardship of the environment and the regional economy to provide a high quality of life for present and future generations. In order for economic development to be considered environmentally sustainable, long-term consequences must be evaluated in terms of impact on the public health and the environment, impact on the economy, and equity for all citizens.
>
> Economic development and environmental protection must be seen as complementary rather than antagonistic processes, both essential to maintain and improve the quality of life of a community through the generations to come.

The key sentence in this passage states: 'Economic development and environmental protection must be seen as complementary rather than antagonistic processes.' It takes a while for it to sink in: these two objectives should work together: they should not be seen as goals in opposition to each other.

Purchasing: the silent partner

Let's face it, purchasing is part of every activity which involves people trying to obtain goods and services of the right quality at the right price. Although it might not be explicitly mentioned in local environmental or social projects as a significant part of their programs, you can be assured that economic considerations were also calculated before the final purchasing decisions were made. Most of the time you only hear of the finalized environmental project, the ecological reserve, the solar panel roof, recycled content plastic lumber boardwalks, natural gas or electric cars. No mention is made of the environmentally responsible purchasing efforts that were made to bring these initiatives to fruition. In the examples highlighted in this chapter, there were leaders and professionals who negotiated and purchased every piece of material used in the projects; individuals made the decision to design, build and purchase. Please consider this collaborative effort when you embark on your environmentally sensitive journey and find and educate your purchasing partners.

Developing local links

Let's take the argument one step at a time. Every good environmental plan or initiative needs to be tested in action, and this applies to environmental purchasing policy as much as to other approaches. And where better to test it than in its local setting? Environmentally sensitive initiatives look wonderful on paper and on the various Web sites now available, but what positive impact are they going to have on society at large? How can you measure their effect? Moreover, how can you make sure these initiatives will endure and be self-sustaining? They will be most likely to endure if they contribute to building up the local economy of suppliers and consumers, creating a virtuous circle of local supply and demand. Can environmentally enhanced contracts and programs assist in local economic development programs and how does that contribute to sustainability? If the contracts are designed with economic development strategies in mind, they can make a real difference.

Local economic development can create local jobs, reducing the impacts of travelling to work. If products are made locally, this reduces the impacts from long distance transportation of finished goods and also provides a potential local market for recyclable waste which can in turn be used to create new and needed products nearby – the closed loop system described so frequently in this book. Just as importantly, local economic development can create sustainable communities which are, in turn, better places to do business.

In most cases, we design and award contracts based on questions such as 'How much is the contract going to cost us?' and 'Do we have the budget to award the contract?' Purchasing professionals are rarely asked to assist in the design of contracts which will be under budget, contain high quality products or services, utilize environmentally sensitive information and technology (both scientific and non-scientific technology) and which will, at the same time, foster local economic development. However, this very ambitious total package can be achieved if it is an integral part of contract design process from the very beginning.

The previous chapters showed how environmentally sensitive contract language can enhance the value and effectiveness of your contracts and increase the co-operation from your corporate partners. But how do we get the proper contract language to enhance local product specification and how can this contract be turned into an actual project?

As discussed before, the whole process should start with the idea that the contract is your 'value implementation tool' – the instrument through which you build in the values of your organization or company. First, you must make these values explicit, that is, put them down on paper for everyone to see.

Local Agenda 21 (LA 21)

One very useful tool for anyone looking for a holistic program is Local Agenda 21 (LA21). For once, all the ingredients for the ideal community which we all want to live in have been listed for us, outlined in the Agenda 21 policy from the UN Rio Earth Summit in 1992, which offered a program for sustainable development for the twenty-first century. LA21 could be adopted by every community in the entire world and it is now being undertaken by many governments in the world, although the UK is by far the most actively engaged in it.

The idea of LA21 is to allow issues concerning communities, and particularly their economic, social and environmental concerns, to be openly discussed by the community in order to identify their own needs and make plans for their own future. In addition, LA21 provides local government with a chance to develop a working partnership with local community groups and others. These partnerships can help facilitate programs that allow communities to agree priority needs and make decisions on future projects to meet these needs.

Since the UK began its ground-breaking work on LA21, there have been numerous initiatives which provide the citizen with access to critical information which will help them assess the effects of possible projects which could be undertaken in their community. These projects include major economic or environmental regeneration programs, which may involve an assessment of the potential effects that the proposed project could have on their way of life. By getting this type of collaborative thinking and participation going via LA21, communities are no longer thinking solely of the economic and job creation benefits that new development could bring. They can start making links between these potential projects and the social and environmental aspects of the community. They can consider whether the proposed use of their resources (land, people and capital) will really benefit the community for the future. The final decisions will be loosely based on my favorite topic, the concept and the technique of 'total life-cycle analysis'. The key way in which LA21 links in to this book is that it brings the entire purchasing and consuming system to the table and exposes it to rigorous critical thinking. This is where and how all the final decisions are weighed with respect to how we will manage our resources and what will be our impact on the environment in which we live.

It will be the community, business and local authorities which will be responsible for the continuity and success of these programs, which are often led by the local authorities. The most important ingredient is a well-developed plan and the continuation of the community meetings, workshops and educational programs which led to the plan throughout the whole implementation process.

WWF UK (World Wide Fund for Nature UK), through its Community Education program, has worked in partnership with many local communities in the UK in an effort to foster the growth of LA21. WWF UK describes its work on LA21 as follows:

> Local authorities and businesses are identified as being central to the process of achieving meaningful local improvements, through Local Agenda 21 strategies, but this process crucially depends upon the participation of individuals in their communities. Although much progress is being made, we have found that organizations wishing to develop more sustainable practice are discovering a need for new skills and techniques in order to generate community participation. We are therefore offering a range of resources and projects that support the

process of community participation in the development of initiatives for sustainable living.

Local Agenda 21 provides an important focus for WWF UK's community-based work. WWF UK has been working with a growing number of local authorities in the UK to support the process of community and business participation in the formulation of local plans to develop and implement strategies encouraged in Local Agenda 21.

WWF UK's Community Education program has more recently developed plans to use environmental contract management and design as part of the new Best Value approach to contract management planned for local authorities in the UK by year 2000. These plans include development work to convert UK local government purchasing and procurement systems from the current Compulsory Competitive Tendering (CCT), in which low cost wins the bid, over to the Best Value purchasing system. The explicit connection between the low-cost-wins strategy and environmental ineffectiveness is clear. By changing to a Best Value system, local authorities will gain the ability to evaluate what is in the best interest of the local community as a condition of contract award status. The opportunity is now present to put environmental criteria into the contract system of one of the UK's largest sectors.

Examples of community links

Few of us fully examine whether or not the community in which we live is sustainable, nor how we might participate in creating sustainable communities. Several programs that I have worked on have involved both the university and the citizens living in the communities that surround it. Several environmental management programs from around the world have also looked at this paradigm of 'the organization merging with community' and the results have been extremely positive.

The challenge for this new vision is to enable local citizens to educate the business community, as well as their local community, about environmental issues and to highlight the role of the purchasing process. As purchasing professionals become more active in local issues, it will show that purchasing does have a multidimensional appeal and can embrace a social agenda.

The following examples give us a glimpse into the future, showing how social and cultural issues could be addressed together, both in the economic and in the environmental decision-making process. They show the broader impact of purchasing decisions, and how society at large can benefit from sustainable purchasing, as well as providing some practical illustrations about how sustainable communities – involving all institutions, business and residents – may work in future.

United States Conference of Mayors: local government in action

The best initiatives take place where you live and work, and when you get the local government involved, you can seize this opportunity and run with it. It is unusual for our local governmental officials to get together and agree on anything, but if there is consensus on environmental issues and from there a clear connection to developing the local economy, our politicians will run, not walk, to make the merger happen.

The US Conference of Mayors is an important example for two reasons. Firstly, United States citizens may not know that this initiative is on the 'lips and minds' of our elected local officials. Secondly, once they know about it, citizens can challenge their own mayors to use this information and make it a foundation of their local government and community.

Some of the best environmental and economic initiatives are instigated by local government leaders setting the stage for some of the most important actions that communities can play a part in. The US Conference of Mayors (USCM) Buy Recycled campaign is one of these initiatives and they describe their work as follows:

The United States Conference of Mayors
LOCAL GOVERNMENT RECYCLING PROCUREMENT PROGRAM – 'BUY RECYCLED' CAMPAIGN

Purpose
To train purchasing officers and recycling co-ordinators on how to buy recycled products, and to implement 'buy recycled' purchasing ordinances. This helps to create markets for the recyclable materials collected by municipal recycling programs.

History
The Buy Recycled Training Institute is a project of the USCM and its affiliate, the Municipal Waste Management Association, and is being conducted in co-operation with the US Environmental Protection Agency.

The USCM project began in 1990 as the Buy Recycled Campaign and has made the transition this year into the Buy Recycled Training Institute.

Activities and deliverable products

1   Conduct intensive training seminars in each of the ten EPA regions. Two full-day seminars will be offered in each region, addressing the need to buy recycled products, state and local laws and implementation, federal programs and national standards, organizational commitment, recycled products and available resources, specification writing, co-operative purchasing, closed loop recycling, buying for waste prevention and other issues.

2   Co-ordinate four Buy Recycled Training Conferences in EPA Region 4. These two-day intensive conferences include an assortment of 15+ expert speakers to present profiles on specific product categories (automotive, construction, office supply, paper, etc.) and federal, state, and local buy-recycled laws and programs.

3   Organize a Buy Recycled Awards Program to recognize cities with the most outstanding buy recycled program.

4   Arrange for cities to receive free samples of recycled products to test in their applications.

5   Conduct comparability studies on recycled products, comparing the economics, quality, performance, environmental impacts of recycled products to their virgin counterparts.

6   Disseminate information on legislative developments affecting local government's ability to expand markets for recycled products.

7   Technical assistance is available to cities in developing recycling procurement programs. The technical assistance includes case studies, specifications, product profiles, information resources on where to find recycled products, a buy recycled training manual, and other programs and initiatives.

The USCM initiative is closely linked with the US Environment Protection Agency (EPA) policies. The EPA has argued that:

If you are not buying recycled, you are not recycling ... Possibly even more important than separating out recyclables for collection, buying recycled content products creates the market demand that can pull these commodities out of the waste stream and into the manufacturing sector. This helps to increase the value of recyclables and diversify the marketplace.

The EPA also runs a parallel Jobs Through Recycling (JTR) program, which is stimulating economic growth and the development of the recycling market by assisting businesses and entrepreneurs to process recycled materials or manufacture recycled content products. The JTR program brings together the economic development and recycling communities through grants, networking, and information sharing. JTR builds recycling expertise within economic development agencies and places business development tools (technical assistance, financing and marketing) in the hands of recycling professionals. By manufacturing new products out of discarded materials, recycling spurs innovation and economic development, creating jobs in the process.

The USCM's Buy Recycled initiative has already been taken a step further: purchasing officials are trained so that they can become discriminating and informed purchasers of recycled products, thus closing the loop (as we have spelled out in earlier chapters). Buying products containing recycled material stimulates the market price for waste materials that can be recycled, and that in turn stimulates the recycling industry that manufactures goods from the waste. The USCM has developed a training initiative to take these ideas forward: the Buy Recycled Training Institute. They describe it as follows:

> The Buy Recycled Training Institute, a program designed to provide information on improving companies' buy recycled initiatives and educate purchasing agents on all aspects of closing the loop ... an overview of the necessary components of any buy recycled initiative; participants discuss the barriers they have encountered when attempting to implement a 'Buy Recycled' initiative in their organizations. Participants take part in a specifications writing exercise designed to teach purchasing employees how to bid out for materials without unintentionally disqualifying recycled content products. The one-day seminars are concluded with speakers representing different recycled content materials – plastics, re-refined oil, glass, paper, and steel.

The need to 'close the loop' on recycled materials is fully recognized – the materials must be collected, then remanufactured, and then of course, rebought. Any weak point in that chain of events and the loop collapses. The USCM recognize this clearly, as their documents show. For example:

Buying Recycled Products: Where Do We Go From Here?
The real benefits from an aggressive office paper recycling program won't be realized until we 'close the loop' with an equally aggressive Buy Recycled program. There are both market and policy obstacles to the effective remanufacture of recyclable materials. You can help overcome these by adhering to the following key elements of a good Buy Recycled program.

1 *Commitment to buy.* Organizations must establish a policy to buy recycled products. This commitment is necessary to provide leadership to
   users and to convince manufacturers that a consistent, long-term
   demand exists so that they can invest in recycling equipment.
2 *Review purchasing specifications.* All specifications must be reviewed
   to eliminate prohibitions or limitations against recycled products.
   In addition, more subtle obstacles to purchasing recycled products,
   such as brightness levels for paper, must be identified and revised.
3 *Establish common definitions and percentages.* Organizations should
   use existing minimum content standards and definitions, such as
   those established by President Clinton (20 percent post-consumer
   by 1995), or the US EPA (currently under review). Manufacturers
   cannot supply different products for each of the fifty states, local
   governments, and thousands of private organizations. By making
   one product, manufacturers can produce shelf items instead of spe-
   cialty items, lowering the cost of production.
4 *Variety of products.* Even though paper makes up the largest percentage of the waste stream, buying recycled paper alone will not solve
   the solid waste problem. Organizations should consider buying a
   variety of recycled products, including: paper, toner cartridges, computer disks, files, packaging and insulation.
5 *Testing products.* When switching to a recycled product, continue to
   ensure that the product meets the functional specifications of the
   equipment it will be used on and its particular end uses. Though recycled products do perform as well as their virgin counterpart, continue
   to perform regular testing and be sure to perform the tests 'blind' so
   that recycled and virgin products can be compared without bias.
6 *Phased-in approach.* Organizations should phase in use of recycled
   products so that users can adjust to the program and manufacturers
   can make long-term capital investments to retool equipment to accept
   recycled materials.
7 *Price.* As recycled products become more prevalent, their prices are
   becoming more competitive with their virgin alternative. On occa-
   sion, recycled products may be more expensive than virgin products
   due to tax policies, price fluctuations, or economies of scale in production or end use. In this case, some organizations use price preferences
   (of 5 percent), life-cycle costing, or use of set-asides (where recycled
   products are purchased separately) to support recycled products. Any
   extra funds spent should be viewed as an investment in market development but are often offset by savings realized from your recycling
   program.

8 *Co-operation among manufacturers, vendors and users.* Organizations must actively solicit and publicize bids from manufacturers and vendors of recycled products. Manufacturers and vendors must provide a wide range of recycled products and let users know about the products.

9 *Co-operative purchasing.* Organizations should join together to buy recycled products. These co-operative purchases expand the volume of products purchased, reduce unit costs of recycled products, help insure availability, and establish common definitions and percentages.

10 *Data.* Organizations should keep good records on buying recycled products and share this information with others. The National Recycling Coalition is currently attempting to collect this data.

11 *Waste reduction and recyclability.* In addition to buying recycled products, organizations should buy products that can be recycled in-house, are more durable, or create less waste in the first place.

12 *Influence.* Use your influence in encouraging others to buy recycled products. In addition to having your own 'buy recycled' program, encourage others that you do business with to follow suit. There is no better way to support recycled products than to require consultants, printers or suppliers to have a 'buy recycled' program.

All communities, US and world-wide, should persuade their local representatives to join them in developing local environmentally responsible programs which have strong purchasing and economic guidelines which could help design and sustain an environmentally responsible purchasing program similar to that established by the US Conference of Mayors.

Sustainable Evansville Project

The Sustainable Evansville Project is the story of a whole community realizing what 'environmental impact' is and deciding to do something about it. In the Evansville program (from which we took the quotation that opens this chapter) the community has an environmental outline, or set of goals, that encompasses their whole community and enshrines what they most value. According to the Sustainable Evansville Project, their initiative has been narrowed down to focus on four principal goals:

1 Teach citizens, elected officials, and public and private organizations the advantages and opportunities offered by environmentally sustainable economic development.

2 Catalyze the development of local, environmentally sustainable industries that use regional waste as a resource and create new jobs.

3 Promote optimal use of energy, land, and raw materials while reducing waste and pollution.
4 Encourage protection and restoration of our natural legacy for future generations.

Look at the order in which the Evansville goals – or values – are listed. First, teach people; second, develop the local economy for local people; third, make optimal use of resources and reduce waste; and fourth, ensure long-term thinking about the environment. People first, environment second. Is that selfish? No, it is simply logical. People are the actors that impact on the environment. Look after them and they will look after the natural world.

The Sustainable Evansville initiative has been highlighted in this chapter because they have pulled together everything that has been discussed in this book so far. One of their most impressive projects was their EcoHouse Project initiative. The EcoHouse concept has long had a fascination for me and for many others, but this EcoHouse had real 'meat' to it. William (Bill) Brown, President of the Board of Directors of Sustainable Evansville, suggests that the real reason for the success story behind the business and contracting process for the EcoHouse was as follows:

> Above all, it shows the power of collaboration with pre-existing organizations. Purchasing for EcoHouse was based, in part, upon student recommendations from their research on the Internet and my experience with green specifications from working with the AIA (American Institute of Architects) Committee on the Environment.

While with the AIA Committee on the Environment, Bill Brown worked on the Greening of the White House and the AIA Environmental Resource Guide:

> The American Institute of Architects' Environmental Resource Guide (AIA/ERG), which is supported in part through a co-operative agreement between AIA and US EPA, consists of a living guidance document distributed through a subscription service that provides important environmental data to architects and other persons involved in the design and construction of buildings and other structures. The AIA/ERG is intended to raise the awareness and understanding of professional architects and support staff regarding the environmental impacts of their materials and design decisions.
>
> The premises of the AIA/ERG are that a sustainable society is not only possible, but necessary, and that architects can play a major role in its creation.

The American Institute of Architects' Environmental Resource Guide covers technology, practice and education, and provides non-technical information regarding environmental considerations relevant to architects. It also covers building materials, and includes data on the environmental and health effects of building materials.

The type of collaborative effort shown in the Evansville EcoHouse project, drawing on the experience and expertise of the community, institutions and individual professionals, is an ideal way of sustaining an innovative and untested program. It is worth quoting the Sustainable Evansville description of the EcoHouse at some length as it brings together many of the elements we have discussed in this book:

Site selection
The ideal site would be one which illustrates the advantages of locating in an existing pedestrian-oriented neighborhood where utility and transportation infrastructure is already in place and frequently-used services and amenities are within walking or biking distance.
Home design
Home design should be marketable, compatible with and comparable to existing neighborhood homes, and responsive to the natural elements of the site (trees, drainage, solar orientation, prevailing wind, and view). Home design should minimize energy and material resource waste while maximizing comfort, durability, marketability, and health and safety.
Waste reduction
Careful design and worker education can help minimize waste throughout all stages of the project. This will save on disposal costs associated with the building of a house and save landfill space.
Recycling
Construction debris generated on site can be recycled for use in the manufacture of new products. This transforms waste into valuable raw material, creating new local industry and new local jobs.
Recycled materials
Using materials made with recycled content uses less energy in the manufacturing process than when products are made from virgin materials; saves valuable natural resources; helps develop the markets for recycled products which creates local manufacturing jobs in businesses that make new products from recycled materials; increases the market value of recycled materials; encourages the collection of recyclable materials, which in turn creates jobs in the local recycling collection and processing industries.

Composting
Household composting will be encouraged through the inclusion of a
backyard composting bin and consumer education.

Non-toxic material and product use
The use of non-toxic materials and products: reduces the potential for
indoor air quality problems; minimizes health risks to builders during
construction; reduces the generation of hazardous wastes and lowers
worker exposure to hazardous materials during manufacturing of the
product.

Other material selection issues
In addition to identifying potential recycled and non-toxic materials,
other material selection issues include local availability (to enhance
local economy and reduce transportation energy), durability and ease of
maintenance; environmental impact of manufacturing process health
and safety, marketability; compatibility with local construction skills
and practices; life-cycle cost; potential for reuse or recycling upon demolition.

Energy efficiency
The use of cost-effective, energy-efficient design and technologies will
save the homeowners' money on utility bills; decrease energy consumption and reduce the pollution associated with energy generation; conserve
non-renewable natural resources; result in a quieter, more comfortable,
better quality, more marketable home. Other energy-efficient designs
and technologies may include upgraded levels of insulation; reduced
air infiltration; low-emission window technology; proper window sizing and location; shading of windows.

Additional thermal mass; selection and proper sizing of heating and
cooling systems; proper duct sealing and duct insulation; design for
natural daylighting and ventilation; water-conserving fixtures; energy-
efficient appliances and lighting; and solar technologies.

Landscaping
Careful landscape design can enhance the energy efficiency of the house
(natural shading, wind barriers); minimize maintenance; reduce the
need for motorized lawn equipment; maximize water absorption and
minimize run-off; and increase marketability. Choosing plants native
or appropriate for the climate will also minimize the need for chemical
pest control, fertilization, and watering.

This is a very substantial example to include in the text of this chapter. But
there is enormous value in these local community initiatives as they demonstrate the linking of issues which affect all purchasing decisions. Look, for

example, at this phrase which clearly shows how to write the development of the local economy into your contracts: 'material selection issues include local availability (to enhance local economy and reduce transportation energy).' Here we see, factored in right from the outset, two important elements: concern for the economic development of local business and avoidance of long-distance transport energy costs and pollution from emissions.

Many different partners (including students) have helped formulate and plan these projects so that they make a concrete expression of these abstract values, articulated in the goals for a sustainable Evansville. Evansville projects 'walk the talk'; they practice what they preach. In pursuing the goals that the Sustainable Evansville community set out for themselves, they have proven that a strong environmental program can (and must) be based on a strong economic foundation, and their achievements are living proof that this powerful combination can be accomplished.

Australian eco-cities

Halifax EcoCity Project

There is also truly remarkable work being done in the Halifax EcoCity Project, Australia, another initiative in a series of community-wide environmental agendas. The Bourne Court Project is a compact and localized pilot, developed within five minutes' walk of the main EcoCity site, to test the various eco-design approaches, technologies, funding and organizational systems for the overall project. It was designed as a pilot development to test systems and processes to eliminate any unknowns and enable the main EcoCity Project to achieve its objectives with the full confidence of all participating partners. At Bourne Court, research is centered around five ecological townhouses which will be located in the center of the city of Adelaide. Their goal is to design-to-trial a number of technologies and design strategies for an ecologically self-sustaining development providing an example and test bed for builders, architects, designers, developers and clients.

The EcoCity's own description of the Bourne Court Project shows the range of its innovative elements:

- It brings together examples of all ecologically sustainable development (ESD) technologies on one small, inner-city site.
- It is unconventional in being designed with a very high thermal mass.
- It is a community-based initiative by a non-profit development co-operative.
- It will be an inner-city grid-connected solar power station.
- It will address all aspects of the federal government's Greenhouse Challenge strategy.

- It is an alternative to computer-dependent 'smart house' technologies. It is designed as a passive climate response development incorporating the kind of very difficult orientations that often arise in dense urban environments and will be trialling location specific, 'dumb' systems of automatic ventilation.
- It will be a model of appropriate high/low technology application. It will use very low energy technologies where appropriate (for example, mud brick garden walls), innovative low tech (such as earth walls and vegetable fiber composite panels), 'middle' tech (such as reinforced concrete floor slabs), and high technology (such as photovoltaic cladding).
- The design was developed using both pencil drafting techniques and low tech recycled cardboard models and high technology advanced CAD (thanks to sponsorship of ArchiCAD program and licenses from Graphisoft).

They outline the innovative techniques and systems that will be used, including:

- *Walls*. Earth technologies (poured earth, rammed earth, pressed earth bricks, mudbricks and possibly cast earth), isochanvre (hemp composite with flax composite for comparison)
- *Roofs*. Glass/amorphous silicon (photovoltaic cladding)
- *Water*. Solar septage greenhouse system (constructed wetlands in a glasshouse for all sewage treatment and full biological recycling on site) dependent on funding grey water treatment (on site); stormwater retention on site; rainwater capture and storage.

All of these strategies, goals and objectives are leading towards the development of multiple programs, namely the EcoCity Project. It is an excellent example of a holistic approach to building and land management and development.

Whyalla's EcoCity Project

By taking advantage of its favorable climatic conditions, the city of Whyalla is stepping into a whole new environmental arena. They describe it as follows:

> Whyalla City Council has adopted a strategic approach to establish the city as a leader in the utilization of sustainable technologies in all aspects of the operation of civic services and the built environment.

This includes:

> ... the creativity and cohesiveness of the local community and Council's unity in its desire to create a new future for the city by broadening its economic base and increasing employment opportunities for young people and residents generally.

Whyalla's EcoCity Project brings together features similar to those developed in the Halifax project, but with their own special local priorities. This is a major project in which the Council is looking to develop vacant land in the city centre ecologically to provide a new urban core for the city and to act as catalyst for the new ecological development for the region. They describe their approach as follows:

> The Council is clearly committed to the principles of Ecologically Sustainable Development (ESD) with a significant resource allocation being made to date, with more planned for the immediate future. They will prove that urban development in the arid zone can be more self-sufficient in its energy and resource needs.
>
> The Whyalla EcoCity project will utilize all aspects of appropriate and sustainable technologies including urban design principles:

- building construction techniques, design factors and materials
- conventional energy conservation and alternative energy generation
- sustainable water use
- reuse of effluent.

Urban Ecology Australia (UEA)

Urban Ecology Australia Inc. (UEA) describe themselves as

> a non-profit community group committed to the evolution of socially vital, economically viable and ecologically sustaining human settlements – ecological cities – through education and example.
>
> UEA is the first and only national organization in Australia with urban environmental issues as its primary focus, and in 1997 was recognized by the federal government as a National Special Interest Organization under the Grants to Voluntary Conservation Organizations (GVCO) program. UEA is an accredited non-government organization with the United Nations. UEA works closely with Ecopolis Pty Ltd. Both Ecopolis and UEA were created to make the vision of ecological cities a reality,

and in 1992 co-initiated the internationally acknowledged Halifax EcoCity Project in Adelaide, South Australia. Based in Adelaide, UEA operates primarily from the Centre for Urban Ecology (CUE).

The UEA program is a breath of fresh air: it shows the power of a simple message and how it can blossom into a community-wide program. They describe their goals and objectives as follows:

UEA Inc. is a community-based, non-profit, educational association with an Australia-wide and international membership. UEA is accredited as an NGO with the United Nations. It has the following objectives and purposes:

(a) To educate, inform and ... exchange information about the evolution of ecologically integrated human settlement
(b) To sponsor, undertake and encourage research into the evolution of ecologically integrated human settlement
(c) To ... transform existing human settlements in the direction of eco-
logical integration, health, and social vitality and equity
(d) To ... build ... new ecologically integrated ... human settlements.

Once again we see human well-being coming first on the agenda: get the human dimension right and the environmental dimension will fall into place. The major thrusts in the UEA strategy are:

- To devise and implement a comprehensive water-recycling plan
- To implement the mandatory controls in development policy to require the installation of solar hot water heaters and passive design features to enhance energy efficiency
- To provide financial incentives for the installation of solar hot water heaters
- To develop an integrated cycle network and linear park
- To utilize the Agenda 21 Local Environment Plan Process
- To initiate public campaigns to disseminate appropriate/sustainable building technologies
- To develop an alternative energy research center.

Conclusion

When I leave my home to go to work, I do not forget my neighbors and my community roots and I hope none of you forgets yours either. I fully realize that my business decisions do have an impact on the local community and on distant communities – maybe yours do too. LA21 and the Australian projects bring both worlds together and the results can be magic, but only if the two can meet somewhere in the middle and begin to understand each other. Use your community as the 'living machine' to create a new future which includes making all purchasing and business decisions part of the initial plan. If you design, plan, meet, discuss, strategize ... and then go and meet with your business leaders and purchasing officials, they will feel left out of the community picture. Include them from the start and everyone can share the good environmental fortunes with the local community.

As we begin to examine what progress could be made if we put these strategies into action, while developing other socially beneficial systems, sustainable systems start to emerge. One area that must be included in whatever plan you or your organization develops is the connection to and enhancement of local and state economic development programs. In addition, if organizations nationally and internationally could develop new technologies and training through the establishment of innovative environmentally sensitive contracts, we could assist local small businesses in the development of new competitively priced environmentally sensitive products.

In short, organizations would develop local and wider markets and demand based on strong environmental policies, and individual purchasing departments would be able to identify local businesses to transfer competitively the technology for the products or services on a long-term contractual basis. Working with local government could make this connection a reality if the community participates. It is a good start that needs tremendous support. The system that you can create will give new local as well as global meaning to the established term the 'closed-loop' system.

Bibliography and General Background Reading

ACBE (1996) *Environmental Reporting and the Financial Sector: An approach to good practice*, HMSO, London

ACCA, Aspinall & Co and BRESCU (1997) *Guide to Environmental and Energy Reporting and Accounting*, ACCA, London

Alabaster, T. and Blair, D.J. (1996) 'Greening the University', in Huckle, J. and Stirling, S. (eds), *Education for Sustainability*, Earthscan, London

Alabaster, T. and Walton, J. (1997) 'Practising what they teach: universities, industry and environmental reporting', in *Environment Information Bulletin* 64, Eclipse Group Ltd, London

Ali Khan, S. (1996) *Environmental Responsibility. A Review of the 1993 Toyne Report*, HMSO, London

Athanasiou, T. (1998) *Divided Planet: The Ecology of Rich and Poor*, Georgia UP, USA

Azzone, G., Noci, G., Manzoni, R. and Welford, R. (1996) 'Defining Environmental Performance Indicators: An integrated framework', in *Business Strategy and the Environment*, 5

Bailey, P. and Farmer, D. (1990). *Purchasing Principles and Management*, Pitman, London

Banking and the Environment. Statement by International Banks on Environment and Sustainable Development

See <http://www.itl.net/features/natwest/statmnt.html> for more details

Barry, A. (1996). 'Buyers Start to Spread the Green Message', in *Purchasing and Supply Management*, February 1996

Barton, H. and Bruder, N. (1995) *A Guide to Local Environmental Auditing*, Earthscan, London

Bell, S. and Morse, S. (1999) *Sustainability Indicators: Measuring the Immeasurable?*, Earthscan, London

Bennet, M. and James, P. (eds) (1998) *The Green Bottom Line: Environmental Accounting for Management Current Practice and Future Trends*, Greenleaf Publishing, Sheffield, UK

Birkin and Woodward (1997) Canadian Institute of Chartered Accountants.

Bishop, J. (1994) *Community Participation in Local Agenda 21*, Local Government Management Board, London

Booth, D.E. (1997) *The Environmental Consequences of Growth*, Routledge, London

Bossel, H. (1998) *Earth at a Crossroads: Paths to a Sustainable Future*, Cambridge University Press, Cambridge, UK

Bromley, D. (1995) *The Handbook of Environmental Economics*, Blackwell, Oxford, UK

Business in the Environment (1996) *The Index of Corporate Environmental Engagement – a survey of the FTSE 100 companies*, Business in the Environment, London

Business in the Environment (BIE), Chartered Institute of Purchasing and Supply (CIPS) and Business in the Community (BITC) (1998) *Buying into a Green Future: Partnerships for Change*, Business in the Environment, London

Butner, S. (1996) 'Using the Internet for Environmental Benchmarking' in the *Seattle Journal of Daily Commerce: Environmental Supplement*, Seattle, USA

Bytheway, A. (1995) *Information in the Supply Chain: Measuring Supply Chain Performance*. Working Paper 1/95, Cranfield School of Management, Cranfield, UK

Cahan, J. and Schweiger, M. (1994) 'Product Life Cycle: The Key to Integrating EHS into Corporate Decision Making and Operations', in *Total Quality Environmental Management*, Winter 1994

Cairncross, F. (1995) *Green, Inc: A Guide to Business and the Environment*, Earthscan, London

Calahan Klein, R., Gavahan, K., Pritchett, T., and Olsen, J. (1997) *Greening the Supply Chain: Benchmarking Leadership Company Efforts to Improve Environmental Performance in the Supply Chain*, Business for Social Responsibility, San Francisco

Canadian Institute of Chartered Accountants (1994) *Reporting on Environmental Performance*, CICA

Canadian Institute of Chartered Accountants (1994) *Full Cost Accounting from an Environmental Perspective*, CICA

Carley, M. and Spapens, P. (1998) *Sharing the World: Sustainable Living and Global Equity in the 21st Century*, Earthscan, London

Carew-Reid, J. et al (1994) *Strategies for Sustainable Development: A Handbook for their planning and implementation*, IUCN/IIED/Earthscan, London

Carley, M. and Christie, I (1998; 2nd edn) *Managing Sustainable Development*, International Union for the Conservation of Nature (IUCN)

Chartered Institute of Purchasing and Supply (CIPS), Business in the Environment (BIE) with KPMG (1994) *Buying into the Environment: Guidelines for Integrating the Environment into Purchasing and Supply*, Business in the Environment, London

Clayton, T. and Radcliffe, N. (1996) *Sustainability: A Systems Approach*, Earthscan, London

Commission of the European Communities (1993) *Towards Sustainability. A European Programme of Policy and Action in Relation to the Environment and Sustainable Development*, Official Journal of the European Communities, C138.17, Brussels, May 1993

Commission of the European Communities (1996) 'Progress from the Commission on the Implementation of the European Community Programme of Policy and Action in Relation to the Environment and Sustainable Development' in *COM* (95) 624 final, CEC Brussels

Commonwealth Environmental Protection Agency (1994) *The Development of Scientific Criteria for Commonwealth Government Purchase of Environmentally Preferred Paper Products*, Canberra: Commonwealth Environmental Protection Agency

Creighton Hammond, S. (1998) *Greening the Ivory Tower: Improving the Environmental Track Record of Universities, Colleges and Other Institutions*, MIT Press, USA

Curran, S. (1998) *The Environment Handbook: A Guide for Business*, HMSO, London

Darde, J.P. (1991) *Valuing the Environment: Six Case Studies*, Earthscan, London

Deloitte and Touche (1996) *Back to Basics – A conceptual framework for environmental accounting*, Deloitte and Touche, London

Delphi Group (1996) *Development for Criteria for Green Procurement: Summary Report*, The Delphi Group for the National Round Table on the Environment and Economy, Ottawa, Canada

Department for Education, Department of the Environment and the Welsh Office (1993) *Environmental Responsibility: An agenda for further and higher education*, HMSO, London

Department of the Environment, Department for Education and Employment (1996) *Environmental Responsibility. A Review of the Toyne Report*, HMSO, London

Department of the Environment (1994) *Sustainable Development: The UK Strategy*, HMSO, London

Department of the Environment (1995) *First Report, British Government Panel for Sustainable Development*, HMSO, London

Department of the Environment (1996) *Indicators of Sustainable Development for the United Kingdom*, Government Statistical Service, HMSO, London

DETR (1998) *Improving Local Services Through Best Value. Consultation document*, Department of the Environment, Transport and the Regions, London, March 1998

Dodds, F. (ed.) (1997) *The Way Forward. Beyond Agenda 21*, Earthscan, London

Eden, S. (1994) 'Using Sustainable Development: The business case', in *Global Environmental Change* Vol 4, No 2, Butterworth-Heinemann, Oxford, UK

Ellram, Lisa M. (1994) Focus Study for the Center for Advanced Purchasing Studies, Tempe, USA

See Web site for full study: <http://www.capsresearch.org/research/focuses/tcm.htm>

Elkington, John (1998) *Cannibals with Forks: The Triple Bottom Line of 21st Century Business*, Capstone Publications, Oxford, UK

Etzioni, A. (1993) *The Spirit of Community*, Simon and Schuster, New York

Forrest, J. (1996) 'Corporations and Sustainable Development' in BWZ – The Online Better World Magazine, Issue 6: Seeking Sustainability, October/November/December 1996. Web site: <http://www.betterworld.com/>

Gray, J. (1998) *False Dawn: The Delusions of Global Capitalism*, Granta Books, London

Green Builder *Sustainable Building Technical Manual*, Green Builder program, Austin, Texas, US. Available on: <http://www.greenbuilder.com/general/BuildingSources.html>

Groundwork (1996). *Purchasing and Sustainability*, Groundwork Foundation, Birmingham, UK

Hanley, N. and Spash, C.L. (1994) *Cost Benefit Analysis and the Environment*, Edward Elgar, Cheltenham, UK

Hass, S.E. (1996) *The Citizen's Handbook of North Carolina's Environmental Information Sources*, Department of City and Regional Planning, University of North Carolina at Chapel Hill. On-line version by Andrew Koebrick, University of North Carolina at Chapel Hill <http://sunsite.unc.edu/hass/home.html>

Hawken, P. (1993) *The Ecology of Commerce: How Business Can Save the Planet*, Weidenfield and Nicholson, London

Hawken, P. (1993) *The Ecology of Commerce. A Declaration of Sustainability*, Harper Collins, New York

Hill, J et al (1994). *Benefiting Business and the Environment: Case studies of cost savings and new opportunities from environmental initiatives*, Institute of Business Ethics, London

Hill, K.E. (1997) 'Supply-Chain Dynamics, Environmental Issues and Manufacturing Firms', in *Environment & Planning* A 29

Hutchinson, C. (1997) *Building to Last: The Challenge for Business Leaders*, Earthscan, London

International Institute of Sustainable Development (IISD) (1993) *Coming Clean: Corporate Environmental Reporting*, Deloitte Touche, Tohmatsu International, London

IUCN/IDRC International Assessment Team (1997) *An Approach to Assessing Progress toward Sustainability*, IUCN

Jacobs, M. (ed.) (1996) *The Politics of the Real World*, Earthscan, London

Jones, C. (1995) *Working in Neighbourhoods: WWF and Local Agenda 21*, WWF UK, Godalming, UK

Jonson, G. (1997) *LCA: A Tool for Measuring Environmental Performance*, Pira International, London

Knight, A. (1995) *How Green Is My Front Door?*, B & Q Environmental Report, Eastleigh, UK

Korten, D. (1995) *When Corporations Rule the World*, Earthscan, London

KPMG (1996) *International Survey of Environmental Reporting*, KPMG, London

KPMG (1997) *The KPMG Survey of Environmental Reporting*, KPMG Summerhouse, London, UK

Lafferty, W.M. and Eckerberg, K. (1998) *From the Earth Summit to Local Agenda 21*, Earthscan, London

Lamming, R. and Hampson, J. (1996) 'The Environment as a Supply Chain Management Issue', in British Journal of Management, special issue.

Levy, S. (ed.) (1998) *Municipal Solid Waste (MSW) Factbook*, US Environmental Protection Agency. Internet Version from <http://www.epa.gov/epaoswer/non-hw/muncpl/factbook/internet>

LGMB (1997) *Annual survey of local authority involvement in Local Agenda 21*, Local Government Management Board, London

McIntyre, K., Smith, H., Henham, A. and Prettore, J. (1998) 'Environmental Performance Indicators for Integrated Supply Chains', in Supply Chain Management.

Meadows, D.H., Meadows, D.L. and Randers, J. (1992) *Beyond the Limits: Global Collapse or Sustainable Future?*, Earthscan, London

Mitchell, J.V. (1998) *Companies in a World of Conflict*, Earthscan, London

Montague, C. (1995) In *American Demographics*, November/ December 1995 issue, Penton Publishing, California

Murphy , D.F. and Bendell, J. (1997) *In the Company of Partners: Business, environment groups and sustainable development post-Rio*, The Policy Press, University of Bristol, Rodney Lodge, Grange Road, Bristol BS8 4EA

Muschett, F.D. (1997) *Principles of Sustainable Development*, St Lucie Press, Florida

Nelder, C. (1996) 'Envisioning a Sustainable Future' in *BWZ – The Online Better World Magazine*, Issue 6: Seeking Sustainability, October/November/December 1996

Odum, H. T. (1996) *Environmental Accounting: Energy and Environmental Decision-Making*, John Wiley, Chichester, UK

OECD (1997) *Economic Globalisation and the Environment*, OECD, Paris, France

OECD (1998) *Towards Sustainable Development: Environmental Indicators*, OECD, Paris, France

Office of the Federal Environment Executive (OFEE) (1998) Executive Order (EO) number 13101 *Greening the Government through Waste Prevention, Recycling and Federal Acquisition*, published by the Office of the Federal Environment

Executive, signed by President Clinton in October 1998; available on <http://www.ofee.gov/guide.htm>

(1998) *Greening the Government. A Guide to Implementing Executive Order 13101*

O'Riordan, T. (1983) *Environmentalism*, Pion Ltd, London

O'Riordan, T. (ed.) *Ecotaxation*, Earthscan, London

O'Riordan, T. and Voisey, H. (1998) *The Transition to Sustainability: The Politics of Agenda 21 in Europe*, Earthscan, London

Piasecki, B.W., Fletcher, K.A. and Mendelson, F.J. (1999) *Environmental Management and Business Strategy*, John Wiley, Chichester, UK

Ponting, C. (1991) *A Green History of the World*, Penguin, London

Rathje, W. and Murphy, C. (1992) *Rubbish! The Archaeology of Garbage: What our garbage tells us about ourselves*, HarperCollins, New York

Richards, R. *Positively Wellington*, Wellington City Council, New Zealand. Web site: <http://www.wcc.govt.nz/wcc/commissioning/economic/strategic_plan/key.shtml>

Rogers, J. J. W. and Feiss, P.G. (1998) *People and the Earth: Basic Issues in Sustainability of Resources and Environment*, Cambridge University Press, Cambridge, UK

Roome, N. J. (ed) (1998) *Sustainability Strategies for Industry: The Future of Corporate Practice*, Island Press, Washington DS

Roseland, M. (1998) *Towards Sustainable Communities. Resources for Citizens and their Governments*, New Society, Gabriola Island, British Columbia, Canada

Russel, T. (1998a) *Sustainable Business: Economic development and environmentally sound technologies*, Greenleaf, Sheffield, UK

Russel, T. (ed.) (1998b) *Greener Purchasing: Opportunities and Innovations.* Includes studies of environmental public procurement at both national and local levels in the USA and in Europe, private enterprise initiatives both in large corporations and in small- and medium-sized enterprises around the globe, including Australia and Japan. Also carries news of green purchasing networks in Japan and Europe operated via the Internet.

Rutgers (1992) Internal memorandum between author and colleagues, Rutgers University.

Rutgers (1992a) *Recycled Products Procurement*, Rutgers University

Rutgers (1992b) *List of Policy Options for the Recycling and Resource Reduction Committee*, Rutgers University

Rutgers (1993) *Recycling, Waste Reduction and Procurement of Recycled Content Products*, Rutgers University Procurement Policies 59201P, 21.1.93

Rutgers (1998) *Rutgers University Purchasing Guidelines and Procedures for Faculty/Staff*, Rutgers University Procurement and Contracting Division, Revised February 1998

Saeed, K. (1998) *Towards Sustainable Development: Essays On System Analysis Of National Policy*, Ashgate Publishing, Aldershot, UK

Sarkar, S. (1998) *Eco-Socialism or Eco-Capitalism: A Critical Analysis of Humanity's Fundamental Choices*, Zed Books, London

Sasseville, D., Wilson, G.W. and Lawson, R. (1997) *ISO 14000 Answer Book: Environmental Management for the World Market*, John Wiley, Chichester, UK

Sayre, D. (1996) *Inside ISO 14000: The competitive advantage of environmental management*, St Lucie Press, Florida

Schwarz, W. and Schwarz, D. (1998) *Living Lightly: Travels in Post-Consumer Society*, Jon Carpenter, Chipping Norton, UK

Schmidheiny, S. (1992) *Changing Course: A Global Business Perspective on Development and the Environment*, MIT Cambridge, Mass. and London

Seager, J. (1995) *The State of the Environment Atlas*, Penguin, London

Sheldon, C. (1997) *ISO 140001 and Beyond: Environmental Management Systems in the Real World*, Greenleaf, Sheffield, UK

Smith, F. (ed.) (1997) *Environmental Sustainability: Practical Global Implications*, St Lucie Press, Florida

Stirling, S. (1998) *WWF and Education for Sustainable Development*, WWF UK, Godalming, UK

Toke, D. (1995) *The Low Cost Planet. Energy and Environmental Problems – Solutions and Costs*, Pluto Press, London

University leaders for a Sustainable Future (ULSF) 'The Declaration', 2100 L Street NW, Washington DC, 20037, USA

UNEP (1995) 'Corporate Environmental Reporting, Programme Summary', *Proceedings of the Corporate Environmental Reporting and Accounting Conference*, SustainAbility, London

United Nations Environment Program (UNEP) (1996) *Engaging Stakeholders*, Vols 1 and 2, SustainAbility, London

United States Environment Protection Agency (USEPA) Office of Solid Waste *Comprehensive Procurement Guidelines*, available on <http://www.epa.gov/epaoswer/non-hw/procure/comp.htm>

US Bureau of Labor Statistics (1998–99) Occupational Outlook Handbook Home Page. <http://stats.bls.gov/oco/ocoso23.htm>

Wackernagel, M. and Rees, W. (1996) *Our Ecological Footprint: Reducing Human Impact on the Earth*, New Society Publishers, British Columbia

Wanless, D. (1995) The New Environmental Agenda – Sustainability and Local Agenda 21. Speech given at the Café Royal, London, March 1995

Warburton, D. (ed.) (1998) *Community and Sustainable Development. Participation in the Future*, Earthscan and WWF UK, London

WCED (World Commission on Environment and Development) (1987) *Our Common Future. The Report of the World Commission on Environment and Development*. Also known as the Brundtland Report. Oxford University Press, Oxford

von Weizsacker, E., Lovins, A., Hunter, L. (1997) *Factor Four: Doubling Wealth – Halving Resource Use*, Earthscan, London

Welford, R. (ed.) (1996) *Corporate Environmental Management 1: Systems and Strategies*, Earthscan, London

Welford, R. (ed.) (1996) *Corporate Environmental Management 2: Systems and Strategies*, Earthscan, London

Willums, J. (1998) *Sustainable Business Challenge: A Briefing for Tomorrow's Business Leaders*, Greenleaf, Sheffield, UK

WWF UK/LGMB (1997 and 1998) *Local Agenda 21 Case Studies of Work for Local Sustainability*, WWF, Godalming, UK

Journals

Business in the Environment News
Kingsway House
103 Kingsway
London WC2B 6QX
Tel: 0171-405 4767; fax: 0171-405 4768

Green Futures
Unit 5 50-56 Wharf Road,
London N1 7SF
Tel: 0171-608 2332; fax: 0171-608 2333

Contacts

ACCA (Association of Chartered and Certified Accountants)
29 Lincoln's Inn Fields,
London WC2A 3EE, UK
Tel: 0171-242 6855; fax: 0171-831 8054

ALFA project
Web site: <http://cei.sunderland.ac.uk/uscei/enved/ALFA/project.htm>

American Demographics/Marketing Tools
Web site: <http://www.marketingtools.com/>
A large Web site with demographic and marketing information, American Demographics supplies access to past issues of American Demographics and Marketing Tools magazines, plus the ability to keyword search each magazine or the entire site. Search for current and archival full-text articles dealing with consumer demographics, marketing issues, statistics, purchasing behavior, and media, lifestyles, and attitudes. Data from *Forecast* magazine also is included, but is only available to subscribers. Any user, however, can join editors and readers in an on-line discussion about demographic data, marketing issues, and more. There is also an on-line catalog for data sources, books, and software, plus access to special data issues, and advertising and subscription information.

The American Institute of Architects (AIA)
Committee on the Environment
1735 New York Ave., N.W.
Washington D.C. 20006
Tel: (202) 626-7300; fax (202) 626-7518

B & Q plc
1 Hampshire Corporate Park
Chandlers Ford, Eastleigh
Hants SO53 3YX, UK
Web site: <http://www.diy.co.uk>

BEES (Building Environmental Education Solutions, Inc.)
685 College Road East
Princeton, New Jersey 08543
Tel: (609) 243-4507; fax (609) 951-8410
Email: kwintress@amre.com
Web site: <http://www.beesinc.org/>

BiE (Business in the Environment)
44 Baker Street,
London W1M 1DH, UK
Tel: 0171-224 1600; fax: 0171-486 1700

Web site: <http://www.bite.org.uk>
or <http://www.business-in-the-environment.org.uk>

CAPS (Center for Advanced Purchasing Studies)
2055 E. Centennial Circle
PO Box 22160
Tempe, AZ 85285-2160, USA
Tel: (602) 752-2277; fax (602) 491-7885
Web site: <http://www.capsresearch.org>
Information and research relating to the role of purchasing in US organizations.

CEED (Community Environmental Education Developments)
Benedict Building, St. George's Way
Sunderland SR1 7BZ, UK
Tel: 0191-515 2184/515 2548; fax: 0191-515 2741
Web sites: <http://www.sunderland.ac.uk/~cs0ceed/>
and <http://www.most.org.pl/CEED/pages.webANN96.htm>

Center for Livable Communities
Offers sustainable land use guidance.
Web site: <http://www.lgc.org/clc/>

Center for Renewable Energy and Sustainable Technology (CREST)
Web site: <http://crest.org/>

Chartered Institute for Purchasing and Supply (CIPS)
Easton House, Easton on the Hill,
Stamford, Lincs PE9 3NZ, UK
Tel: 01780 756 777; fax: 01780 756 610
Web site: <www.cips.org>
The professional association for purchasing in the UK.

Department of the Environment, Transport and the Regions (DETR)
Eland House, Bressenden Place,
London SW1E 5DU
Tel: 0171-890 3333
Web site: <http://www.detr.gov.uk>

DTTI (Deloitte and Touche)
Orlyplein 50, NL-1043 DP,
Amsterdam, The Netherlands
Tel: +33 31 20 6 06 11 00
Fax: +33 31 20 6 81 12 60

Earthscan Publications Ltd
120 Pentonville Road
London N1J 9JN, UK
Tel: 0171-278 0433; fax: 0171-278 1142

Environmental Education and Environmental Reporting Network
Web site: <http://cei.sunderland.ac.uk/international.clearinghouse/index.html>
or <http://cei.sunderland.ac.uk/uscei/envrep.index.htm>

Ethical Investment Research and Information Service (EIRIS)
Tel: 0171-735 1351

Federal Environmental Protection Agency
Contact the FEPA or regional government office for official guidelines and procedures.
Web site: <http://www.epa.gov or www.state.nj.us/dep.>

Forum for the Future
Thornbury House,
18 High Street,
Cheltenham, Glos GL50 1DZ
or
Forum for the Future
Sustainable Economics Unit and Best Practice Directory
227a City Road,
London EC1V 1JT, UK
Tel: 0171-251 6070; fax: 0171-251 6268

GreenDisk, Inc.
Web site: <http://www.greendisk.com>

Green Shopping
Your place to shop for eco-friendly products and information from ethical manufacturers with a commitment to the environment.
Web site: <http://www.greenshopping.com/tw/greenshopping/>

Greenleaf Publishing
Broom Hall, 8-10 Broomhall Road,
Sheffield S10 2DR, UK
Tel: 0114-266 3789; fax: 0114-267 9403
Email: greenleaf@worldscape
Web site: <http://www.greenleaf-publishing.com>

Halifax EcoCity Project, Australia
Web site: <http://www.eastend.com.au/~ecology/halifax/bourne.html>

Hewlett Packard
Web site: <http://www.hp.com>

Institute for Global Futures Research Global Futures Bulletin
Email publication at igfr@peg.apec.org

International Institute for Environment and Development (IIED)
3 Endsleigh Street

London WC1H ODD, UK
Web site: <www:oneworld.org/iied>

IISD
161 Portage Avenue East
Winnipeg, Manitoba R3B 0Y4
Canada
Tel: (204) 958 7700; fax: (204) 958 7710

International Purchasing and Supply Education and Research (IPSERA)
c/o Professor Richard Lamming
Chairman IPSERA
CIPS Chair of Purchasing and Supply,
University of Bath School of Management
Claverton Down, Bath BA2 7AY, UK
Tel 01225 826536; fax 01225 826210
A unique, active, international network of academics and practitioners dedicated
to the development of understanding on matters concerning the future of pur-
chasing and supply management.

Dr Thomas Kelly
Director, Sustainability Programs
University of New Hampshire
206-A Nesmith Hall
Durham, NH 03824 USA
Tel: (603) 862-2640; fax (603) 862-0785
Email: thkelly@hopper.unh.edu

KPMG
PO Box 695, 8 Salisbury Square,
London EC4Y 8BB, UK
Tel: 0171-311 1000; fax: 0171-311 3311

Massachusetts Institute of Technology (MIT)
Web site: <http://web.mit.edu/ctpid/www/tbe/overview.html>

National Association of Educational Buyers, Inc.
450 Wireless Boulevard w Hauppauge,
NY 11788, USA
Tel: (516) 273-2600; fax: (516) 952-3660
Web site: <www.naeb.org>
NAEB is the professional association serving colleges and universities nationwide.
It was launched in 1921 to provide a forum where campus purchasing professionals might share valuable information. The professional association dedicated
to serving higher education by providing those with purchasing responsibility with
the specialized information they must have. Their mission is to advocate the
development, exchange and practice of effective and ethical procurement principles and techniques within the higher education community, through continuing
education, publications and networking opportunities.

National Association of Purchasing Management (NAPM)
PO Box 22160
Tempe, AZ 85285-2160, USA
Tel: 800-888-6276
Web site: <www.napm.org>

National Institute of Governmental Purchasing, Inc.
11800 Sunrise Valley Drive, Suite 1050
Reston, Virginia 20191-5303, USA
Tel: 703-715-9400; fax: 703-715-9897
Web site: <www.nipg.org>

National Wildlife Federation's Campus Ecology Program (NWF/CE)
Julian Keniry, National Co-ordinator
NWF, Campus Ecology Program
8925 Leesburg Pike
Vienna, VA 22184
Tel: 703/790-4322
Email: jkeniry@nwf.org
Resources include Ecodemia, Campus Ecology Tool Kit and National Training
Clinics (an intensive, interactive five-day training for student leaders on campus
greening is designed to give participants a head start in launching projects for the
academic school year).

Natural History Book Service (NHBS) Mailorder Books
2-3 Wills Road, Totnes,
Devon TQ9 5XN, UK
Tel: 01803 865 913; fax: 01803 865 280
orders@nhbs.co.uk
Web site: <www.nhbs.com>
NHBS Alert: free email update of environmental books available: register your email
address with NHBS via their web site.

New Economics Foundation
Cinnamon House, 6-8 Cole Street,
London SE1 4YH
Tel: 0171-407 7447; fax: 0171-407 6473
Email: neweconomics@gn.apc.org

Plastics Resource
Web site: <http://205.231.87.246/Docs/apc/>
The information in this site has been organized and formatted to meet the needs
of specific user groups.

Recycled Pulp and Paper Coalition
Web site: <http://www.recycledpulp.com/>
The basics about post-consumer recycled paper guide to purchasing and using
recycled content paper.

Recycler's World
Web site: <http://www.recycle.net/>
Recycler's World was established as a world-wide trading site for information related to secondary or recyclable commodities, by-products, used and surplus items or materials.

Somat
855 Fox Chase
Coatesville, PA 19320 USA
Tel: 610-384-7000; fax: 610-380-8500

Steelcase
Web site: <http://www.steelcase.com>

Sustainable community
Web sites can be accessed via <http://www.geonetwork.org/>

Sustainable Design
Architecture School, Ball State University, Muncie, Indiana.
Web site: <http://www.bsu.edu/cap/arch/arch.html>

Sustainable Evansville Project
Web site: <http://www.sustainableevansville.org/>

Thomas Register of American Manufacturers
Web site: <http://noframes.thomasregister.com/index.html>
Provides access to a search engine which allows you to search the 155,000 companies in our database classified under more than 57,000 product and service headings. Plus, more than 3,100 on-line supplier catalogs give you 44,000 pages of detailed buying and specifying information.

ULSF
2100 L Street NW
Washington, DC 20037, USA
Phone: (202) 778-6133; fax (202) 778-6138
Email: ULSF@aol.com
Web site: <www.ulsf.org>

Urban Ecology Australia Inc.
For more details, see <http://www.eastend.com.au/~ecology/>

US Conference of Mayors (USCM)
J. Thomas Cochran, Executive Director
1620 Eye Street, NW, Washington, DC 20006
Telephone (202) 293-7330; fax (202) 293-2352
Web site: <http://www.usmayors.org/USCM/recycle/buy/brtigi.htm>
Buy Recycled Training Institute: Tel: (202) 822-9058

US Department of Energy Building Energy Tools Directory:
Web site: <http://www.eren.doe.gov/buildings/tools_directory/>

US Department of Energy's Center of Excellence for Sustainable Development
Web site: <http://www.sustainable.doe.gov/>

US Department of Energy's energy efficiency programs
Web site: <http://www.eren.doe.gov/>

US Environmental Protection Agency
Buy Recycled material:
Web site: <http://epainotes1.rtpnc.epa.gov:7777/r10/owcm.nsf/recycle/>

US Environmental Protection Agency
Office of Solid Waste and Comprehensive Procurement Guidelines.
Web site: <http://www.epa.gov/epaoswer/non-hw/procure/comp.htm>

Mathis Wackernagel PhD
Center for Sustainability Studies / Centro de Estudios para la Sustentabilidad
Universidad An‡huac de Xalapa
Apartado Postal 653
91000 Xalapa, Ver. MEXICO
Tel. 28 14-96-11; fax: 28 19-04-53
Web site: <http://www.edg.net.mx/~mathiswa>

Whyalla EcoCity, Australia
Web site: <http://www.eastend.com.au/~ecology/whyalla/brief.html>

WWF-UK
Panda House, Weyside Park
Catteshall Lane
Godalming, Surrey GU7 1XR
Tel: 01483 426444; fax: 01483 426409
Web site: <http://www.wwf-uk.org/education/index.htm>
WWF UK's higher education work Web sites include: <http://www.cecs.ed.ac.uk/greeninfo/GCPACK/Extrat5.htm> or <http://greeninfo.ucs.ed.ac.uk/reports/ecw/paper2.htm>

Yahoo! Yellow Pages
Web sites: <http://yp.yahoo.com/yt.hm?CMD=FILL&SEC=startfind&FAM=yahoo&FT=2&FS=509312&what=reycling+centers>
Find the recycling centers near you. Just enter your street address or intersection, state and zip. This program provides directions and a local map to the facilities.
Multiple resources: recycling, indices, companies, composting, organizations, publications available on Web site: <http://www.yahoo.com/Society_and_ Culture/ Environment_and_Nature/Recycling/>

Index

Index compiled by Sue Carlton